T0183186

SpringerBriefs in Applied Sciences and Technology

Computational Intelligence

Series editor

Janusz Kacprzyk, Warsaw, Poland

For further volumes:
http://www.springer.com/series/10618

Leslie Astudillo · Patricia Melin · Oscar Castillo

Chemical Optimization Algorithm for Fuzzy Controller Design

 Springer

Leslie Astudillo
Patricia Melin
Oscar Castillo
Division of Graduate Studies
Tijuana Institute of Technology
Tijuana
Mexico

ISSN 2191-530X   ISSN 2191-5318   (electronic)
ISBN 978-3-319-05244-1 ISBN 978-3-319-05245-8   (eBook)
DOI 10.1007/978-3-319-05245-8
Springer Cham Heidelberg New York Dordrecht London

Library of Congress Control Number: 2014933533

Printed on acid-free paper

Springer is part of Springer Science+Business Media (www.springer.com)

Preface

In this book, a novel optimization method inspired by a paradigm from nature is introduced. The chemical reactions are used as a paradigm to propose an optimization method that simulates these natural processes. The proposed algorithm is described in detail and then a set of typical complex benchmark functions is used to evaluate the performance of the algorithm. Simulation results show that the proposed optimization algorithm can outperform other methods in a set of benchmark functions.

This chemical reaction optimization paradigm is also applied to solve the tracking problem for the dynamic model of a unicycle mobile robot by integrating a kinematic and a torque controller based on fuzzy logic theory. Computer simulations are presented confirming that this optimization paradigm is able to outperform other optimization techniques applied to this particular robot application.

This book is intended to be a reference for scientists and engineers interested in applying optimization techniques for solving problems in pattern recognition, intelligent control, intelligent manufacturing, robotics, and automation. This book can also be used as a reference for graduate courses like the following: soft computing, intelligent pattern recognition, computer vision, applied artificial intelligence, and similar ones. We consider that this book can also be used to get novel ideas for new lines of research, or to continue the lines of research proposed by the authors of the book.

In Chap. 1, we begin by offering a brief introduction to the potential use of the chemical optimization method in different real-world applications. We describe the use of this method for optimizing type-2 fuzzy systems in problems of intelligent control of nonlinear plants. We also mention other possible applications of the proposed chemical optimization paradigm.

We describe in Chap. 2 the basic concepts, notation, and theory of nature-inspired optimization in general, and then of chemical optimization, in particular. This chapter overviews the background, main definitions, and basic concepts, useful for the development of this investigation work.

We describe in Chap. 3 a brief overview of the basic concepts from chemical methods needed for this work. In this chapter, some definitions of the chemical reactions are offered as a basis for understanding the proposed chemical optimization algorithm.

Chapter 4 introduces the chemical reaction algorithm and describes the main characteristics and definitions. Once the optimization algorithm is presented, it is applied to a set of benchmark functions; the obtained results are compared with the results obtained with genetic algorithms and other optimization paradigms. In the second stage, this chemical reaction algorithm is applied to optimize the parameters of the membership functions of a Type-1 and Type-2 fuzzy logic controller for the tracking of a unicycle mobile robot. With this work, we demonstrate that this novel optimization paradigm is able to perform well in these specific topics and will encourage its use in different soft computing approaches.

We describe in Chap. 5 the application problems used to test the proposed chemical method. In this chapter, first a set of benchmark mathematical functions are presented, which are standard functions usually considered to test new optimization algorithms. In a second phase a more complicated problem of designing type-1 and type-2 fuzzy controllers for an autonomous mobile robot is also considered as a test for the chemical optimization paradigm.

We describe in Chap. 6 the simulation results and comparison with other approaches for both kinds of application problems presented in Chap. 5. First, the simulation results for a set of benchmark mathematical functions are presented. Second, the simulation results for the problem of designing type-1 and type-2 fuzzy controllers for an autonomous mobile robot is presented and compared against other methods, like genetic algorithms.

We offer in Chap. 7 the conclusions of this work, and envision some future research work. Basically, a new optimization method based on mimicking the chemical reactions in nature was introduced. The main characteristics of this algorithm are the exploiting mechanisms combined with the elitist survival strategy, which prevents the algorithm from stagnating in local optima.

We end this Preface of the book by giving thanks to all the people who have helped or encouraged us during the writing of this book. First of all, we would like to thank our colleague and friend Prof. Janusz Kacprzyk for always supporting our work, and for motivating us to write our research work. We would also like to thank our colleagues working in Soft Computing, who are too many to mention by name. Of course, we must thank our supporting agencies, CONACYT and DGEST, in our country for their help during this project. We have to thank our institution, Tijuana Institute of Technology, for always supporting our projects. Finally, we thank our families for their continuous support during the time that we spent in this project.

Tijuana, Mexico   Leslie Astudillo
                                                                   Patricia Melin
                                                                   Oscar Castillo

Contents

Chapter 1
Introduction

Abstract This chapter offers an introduction to the optimization method based on the paradigm of chemical reactions and its application to the design of fuzzy controllers in robotic systems.

Keywords Chemical optimization • Robot control • Chemical reactions • Optimization

Optimization is an activity carried out in almost every aspect of our life, from planning the best route in our way back home from work to more sophisticated approximations at the stock market, or the parameter optimization for a wave solder process used in a printed circuit board assembly manufacturer optimization theory has gained importance over the last decades. From science to applied engineering (to name a few), there is always something to optimize and of course, more than one way to do it.

In a generic definition, we may say that optimization aims to find the "best" available solution among a set of potential solutions in a defined search space. For almost every problem exists a solution, not necessarily the best, but we can always find an approximation to the "ideal solution", and while in some cases or processes is still common to use our own experience to qualify a process, a part of the research community have dedicated a considerably amount of time and efforts to help find robust optimization methods for optima finding in a vast range of applications.

It has been stated the difficulty to solve different problems by applying the same methodology, and even the most robust optimization approaches may be outperformed by other optimization techniques depending on the problem to solve.

When the complexity and the dimension of the search space make a problem unsolvable by a deterministic algorithm, probabilistic algorithms deal with this problem by going through a diverse set of possible solutions or candidate solutions. Many metaheuristic algorithms can be considered probabilistic, while they apply probability tools to solve a problem, metaheuristic algorithms seek good solutions by mimicking natural processes or paradigms. Most of these novel optimization paradigms inspired by nature were conceived by merely observation of an existing process and their main characteristics were embodied as computational algorithms.

The importance of the optimization theory and its application has grown in the past few decades, from the well known Genetic Algorithm paradigm to PSO, ACO, Harmonic Search, DNA Computing, among others, they all were introduced with the expectation of improving the results obtained with the existing strategies.

Over the past years, there has been a growing interest in solving optimization problems by means of algorithms inspired on natural paradigms [1–4]. These techniques have been applied to the optimization of complex computational problems including forecasting [5], control [6, 7], pattern recognition [8, 9] and trajectory planning [10, 11] among others, and have demonstrated not only to comply with their objectives, but they also promote the creation of new ways to give solutions to these complex problems and improve the actual methods as well [12–15].

When applying intelligent methodologies such as fuzzy systems, neural networks, ANFIS, etc., one of the main difficulties is the tuning of their parameters, which are the key to the success of these methodologies. Their parameters vary depending on the complexity of the problem and the method used to find the solution; and in some cases, they stem from our own ability to conceptualize the problem itself, taking into account, the inputs of the system and the expected output values.

In this research work, the main objective is to introduce a new optimization algorithm based on the chemical reactions, the process in which the different elements existing in nature are created, behave and interact with each other to form chemical compounds.

This Chemical Reaction Algorithm (CRA) was applied to diverse optimization problems to prove its performance.

References

1. L.M. Adleman, Molecular computation of solutions to combinatorial problems. Science **266**(5187), 1021–1024 (1994)
2. M. Dorigo, V. Maniezzo, A. Colorni, Ant system: optimization of cooperating agents. IEEE Syst. Man Cybern. Soc. **26**(1), 29–41 (1996)
3. R.C. Eberhart, J. Kennedy, A new optimizer using particle swarm theory, in *Proceedings of 6th International Symposium on Micro Machine and Human Science*, Nagoya, Japan, 1995, pp. 39–43
4. J.H. Holland, Genetic algorithms. Sci. Am. **267**(1), 44–50 (1992)
5. R. Araújo, A. de, Swarm-based translation-invariant morphological prediction method for financial time series forecasting. Inf. Sci. **180**, 4784–4805 (2010)
6. R. Martinez, O. Castillo, L. Aguilar, Optimization of type-2 fuzzy logic controllers for a perturbed autonomous wheeled mobile robot using genetic algorithms. Inf. Sci. **179–13**, 2158–2174 (2009)
7. A. Melendez, O. Castillo, Genetic algorithm with neuro-fuzzy fitness function for optimal fuzzy controller design, in *Proceedings of Fuzzy Information Processing Society, NAFIPS'11*, El Paso, TX, USA, 2011, pp. 1–5
8. D. Hidalgo, O. Castillo, P. Melin, Type-1 and type-2 fuzzy inference systems as integration methods in modular neural networks for multimodal biometry and its optimization with genetic algorithms. Inf. Sci. **179**, 2123–2145 (2009)

9. O. Mendoza, P. Melin, G. Licea, A hybrid approach for image recognition combining type-2 fuzzy logic, modular neural networks and the Sugeno integral. Inf. Sci. **179**, 2078–2101 (2009)
10. R.J. Duro, M. Graña, J. de Lope, On the potential contributions of hybrid intelligent approaches to Multicomponent Robotic System development. Inf. Sci. **180**, 2635–2648 (2010)
11. G. Ma, H. Duan, S. Liu, Improved ant colony algorithm for global optimal trajectory planning of UAV under complex environment. Int. J. Comput. Sci. Appl. **4–3**, 57–68 (2007)
12. S. Gao, Z. Tang, H. Dai, J. Zhang, A hybrid clonal selection algorithm. Int. J. Innovative Comput. Inf. Control **4**(4), 995–1008 (2008)
13. O. Montiel, O. Castillo, P. Melin, A. Rodriguez Diaz, R. Sepulveda, Human evolutionary model: a new approach to optimization. Inf. Sci. **177**, 2075–2098 (2007)
14. E. Rashedi, H. Nezamabadi-pour, S. Saryazdi, GSA: a gravitational search algorithm. Inf. Sci. **179**, 2232–2248 (2009)
15. X. Wang, X.Z. Gao, S.J. Ovaska, A hybrid optimization algorithm based on ant colony and immune principles. Int. J. Comput. Sci. Appl. **4**(3), 30–44 (2007)

Chapter 2
Theory and Background

Abstract This chapter describes the theoretical concepts and background required to understand the new nature inspired optimization method based on the paradigm of chemical reactions.

Keywords Chemical reactions • Chemistry • Nature inspired optimization

This chapter overviews the background and main definitions and basic concepts, useful to the development of this investigation work.

2.1 Nature-Inspired Metaheuristics

The importance of dealing with optimization theory has grown due to the large variety of fields where optimization is applied (mathematics, computer science, engineering, etc.).

For most problems (of mathematical nature), there is more than one path to arrive to a correct solution and that's where the optimization algorithms come into play. As mentioned before, for some problems, deterministic algorithms are not suitable, due the complexity and/or large dimension of the search space; probabilistic algorithms on the other hand, deal with this problem by going through a diverse set of possible solutions or candidate solutions.

Many metaheuristic algorithms can be considered probabilistic, while they apply probability tools to solve a problem, metaheuristic algorithms seek good solutions by mimicking natural processes or paradigms.

Most of these novel optimization paradigms inspired by nature were conceived by merely observation of an existing process. These insights were embodied as computational algorithms and were later applied to solve diverse kinds of problems.

L. Astudillo et al., *Chemical Optimization Algorithm for Fuzzy Controller Design*,
SpringerBriefs in Computational Intelligence, DOI: 10.1007/978-3-319-05245-8_2,
© The Author(s) 2014

The advantage when doing this is that –at least at the beginning-, no mathematical equations are involved. It's later, -when these techniques are widely explored-, that the theory, concepts and applications can be widely described.

Some examples of these algorithms are the Genetic Algorithms [1], Ant Colony Optimization [2], Particle Swarm Optimization [3], DNA computing [4], among others.

A *genetic algorithm* is a stochastic global search method that mimics the metaphor of natural biological evolution. This Darwinian evolution theory is a well known paradigm that has been proved to be robust when applied to search and optimization problems [5]. Evolution is determined by a natural selection of individuals (based on their fitness); which, is expected to be better throughout a determined number of generations by means of recombination and mutation operations.

Simulated annealing is a probabilistic metaheuristic algorithm that imitates the annealing process in metallurgy, a technique that involves the slow cooling of a physical system to find low-energy states [6].

Particle Swarm Optimization, was introduced by Kennedy and Eberhart [3]. As it name implies, it was inspired by the movement and intelligence of swarms. A swarm is a structured collection of interacting organisms such as bees, ants, or birds. Each organism in a swarm is a particle or agent. Particles and swarms in PSO are equivalent to individuals and populations in other evolutionary algorithms. The position, or site, of a particle in a swarms is the vector of optimizations parameters x in PSO. Particles in a swarm cooperate by sharing knowledge. This has been shown to be critical idea behind the success of the particle PSO algorithm.

DNA computing, introduced by Adleman [4], appeared in 1994 as a form of computation whose main characteristic is the use of molecular biology instead of the traditional silicon-based computer. In DNA computing, the information is represented by sequences of bases in DNA molecules called *strings* and each molecule encodes a potential solution to the problem. The main advantage when using this optimization paradigm is the inherent parallelism.

The operations that can be applied in this methodology are: synthesis, denaturing, annealing and ligation.

2.2 Artificial Chemistry

2.2.1 The Set of Molecules S

The set of molecules $S = \{s_1, \ldots s_i, \ldots s_n\}$, where n might be infinite, describes all valid molecules that may appear in an artificial chemistry. A molecule's representation is often referred to as its structure and is set in contrast to its function, which is given by the reaction rules R. The description of valid molecules and their structure is usually the first step in the definition of an AC. This step is analogous to the part of chemistry that describes what kind of atomic configurations form stable molecules and how these molecules appear.

2.2.2   The Set of Rules R

The set of reaction rules R describes the interactions between molecules. A rule can be written according to the chemical notation of reaction rules in the form:

$$s_1 + s_2 + \ldots s_n \tag{2.1}$$

A reaction rule determines the n components (objects, molecules) on the left-hand side that can react and subsequently be replaced by the m components on the right-hand side. n may be called the *order* o the reaction. Note that the "+" sign is not necessarily an operator here, but only separates the components on either side.

2.2.3   Reactor Algorithm A–Dynamics

An algorithm determines how the set R of rules is applied to a collection of molecules P, called *reactor*; *soup, reaction vessel or Pool*. Note that P cannot be identical to S since some molecules might be present in many exemplars, others not at all.

Algorithm A depends on the representation of P. In the simplest case, without a spatial structure in P, the collection of molecules can be represented explicitly as a multi set or implicitly as a concentration vector.

2.3   Fuzzy Logic

Zadeh introduced the term fuzzy logic in his work "fuzzy sets", where he described the mathematics of the fuzzy set theory in 1965.

Fuzzy logic gives the opportunity to model conditions that are defined with imprecision.

The tolerance of the fuzzy in the process of human rezoning suggests that most of the logic behind the human rezoning is not the traditional bi-valued logic, or even the multi-valued, but the logic with fuzzy values, with fuzzy connections and fuzzy rules or inferences.

2.3.1   Fuzzy Sets

Fuzzy sets are an extension of the classic set theory and, as it name implies it, it is a set with boundaries not well defined, this means that the transition of belonging or not belonging to certain set is gradual, and this smooth transition is characterized by grades of membership that gives the fuzzy sets flexibility in modeling linguistic expressions commonly used, such as "the weather is cold" or "Gustavo is tall".

2.3.2 Fuzzy Logic Controller

Fuzzy control is a control method based on fuzzy logic. Just as fuzzy logic can be described simply as "computing with words rather than numbers" fuzzy control can be described simply as "control with sentences rather than equations".

The collection of rules is called a *rule base*. The rules are in the familiar if–then format, and formally the "if" side is called the *antecedent* and the "then" side is called the *consequent*.

Fuzzy controllers are being used in various control schemes; the most used is the *direct control*, where the fuzzy controller is in the forward path in a feedback control system. The process output is compares with a reference, and if there is a deviation, the controller takes action according to the control strategy.

In a *feed forward control* a measurable disturbance is being compensated, it requires a good model, but if a mathematical model is difficult or expensive to obtain, a fuzzy model may be useful. Fuzzy rules are also used to correct tuning parameters. If a nonlinear plant changes operating point it may be possible to change the parameters of the controller according to each operating point. This is called *gain scheduling* since it was originally used to change process gains.

A gain scheduling controller contains a linear controller whose parameters are changed as a function of the operating point in a preprogrammed way. It requires thorough knowledge of the plant, but it is often a good way to compensate for nonlinearities and parameter variations. Sensor measurements are used as *scheduling variables* that govern the change of the controller parameters, often by means of a table look-up.

2.4 Related Work

Optimization based on chemical processes is a growing field that has been satisfactorily applied to several problems.

Artificial chemistry algorithms intend to mimic as close as possible a real chemistry process, by assigning kinetic coefficients, defining molecule representation and focusing on an efficient energy conservation state.

The main difference between these metaheuristics is the parameter representation, which can be explicit or implicit.

A review of scientific work in artificial chemistry can be found in [7]. Chemical inspired paradigms can be differenced by their parameter representation, which can be explicit or implicit. A DNA based algorithm is applied in [8] to solve the small hitting set problem. This NP-complete problem takes exponential time to solve it and it was demonstrated that when using DNA-based supercomputing, only polynomial time is needed to solve it.

In [9] a catalytic search algorithm is explored, where some physical laws such as mass and energy conservation are taken into account. The disadvantage of this algorithm is its slow growth rates and weak selection pressure.

In [10], the potential roles of energy in algorithmic chemistries are illustrated. An energy framework is introduced, which keeps the molecules within a reasonable length bounds, allowing the algorithm to behave thermodynamically and kinetically similar to real chemistry.

A chemical reaction optimization was applied to the grid scheduling problem in [11], where molecules interact with each other aiming to reach the minimum state of free potential and kinetic energies.

The main difference of the proposed chemical optimization algorithm with respect to the previous mentioned approaches is that it has a simpler parameter representation and yet it is proven to be efficient. Since only the general schema of the chemical reactions is taken into consideration, the initial set of elements is simply represented and no extra parameters (such as mass, kinetic coefficient, etc.) are added.

The selected elements react to obtain new compounds and vice versa, and as mentioned before, no previous validation is performed; this improves efficiently the execution rate of the algorithm and ensures a broad exploration of the search space in every iteration, otherwise if the selected parents don't comply with the validation rules then no new products are generated, reducing the possibilities of fully exploring the complete search space.

References

1. J.H. Holland, Genetic algorithms. Sci. Am. **267**(1), 44–50 (1992)
2. M. Dorigo, V. Maniezzo, A. Colorni, Ant system: optimization of cooperating agents. IEEE Syst. Man Cybern. Soc. **26**(1), 29–41 (1996)
3. J. Kennedy, R.C. Eberhart, Particle swarm optimization, in *Proceedings of IEEE International Conference on Neural Networks* (Perth, WA, Australia, 1995), pp. 1942–1948
4. L.M. Adleman, Molecular computation of solutions to combinatorial problems. Science **266**(5187), 1021–1024 (1994)
5. L. Astudillo, O. Castillo, L. Aguilar, Intelligent control for a perturbed autonomous wheeled mobile robot: a type-2 fuzzy logic approach. Nonlinear Stud. **14**(1), 37–48 (2007)
6. P. Salomon, P. Sibani, R. Frost, *Facts, conjectures, and improvements for simulated annealing* (Society for Industrial and Applied Mathematics (SIAM), Philadelphia, PA, 2002)
7. P. Dittrich, J. Ziegler, W. Banzhaf, Artificial chemistries-a review. Artif. Life **7**, 225–275 (2001)
8. N.-Y. Shi, C.-P. Chu, A molecular solution to the hitting-set problem in DNA-based supercomputing. Inf. Sci. **180**, 1010–1019 (2010)
9. Z.W. Geem, Novel derivative of harmony search algorithm for discrete design variables. Appl. Math. Comput. **199**(1), 223–230 (2008)
10. T. Meyer, L. Yamamoto, W. Banzhaf, C. Tschudin, Elongation Control in an Algorithmic Chemistry, Advances in Artificial Life: Darwin Meets von Neumann. Lecture Notes in Computer Science, vol. 5777 (2011), pp. 273–280
11. J. Xu, A.Y.S. Lam, V.O.K. Li, Chemical reaction optimization for the grid scheduling problem. IEE Commun. Soc. ICC **2010**, 1–5 (2010)

Chapter 3
Chemical Definitions

Abstract This chapter describes the basic chemical definitions needed to understand the proposed optimization method based on the paradigm of chemical reactions. The four basic chemical reactions are presented and explained.

Keywords Chemical reactions • Chemical definitions • Elements • Compounds

In this chapter, some definitions of the chemical reactions are offered as a basis for understanding the proposed chemical optimization algorithm.

3.1 Chemical Definitions

Chemistry is the study of matter and energy and the interaction between them, including the composition, the properties, the structure, the changes which it undergoes, and the laws governing those changes. A substance is a form of matter that has a defined composition and characteristic properties. There are two kinds of substances: elements and compounds.

An *element* is a substance that cannot be broken down into simpler substances by ordinary means. Atoms of most of the elements may interact with each other to form compounds. It is apparent from the wide variety of different materials in the world that there are a great many ways to combine elements.

Compounds are substances formed by two or more elements combined in definite proportions through a chemical reaction. There are millions of known compounds, and thousands of new ones are discovered or synthesized each year. A chemical reaction is a process in which at least one substance changes its composition and its sets of properties; they are classified into 4 types [1].

Type 1: combination reactions $(B + C \rightarrow BC)$. A combination reaction is a reaction of two reactants to produce one product. The simplest combination

L. Astudillo et al., *Chemical Optimization Algorithm for Fuzzy Controller Design*,
SpringerBriefs in Computational Intelligence, DOI: 10.1007/978-3-319-05245-8_3,
© The Author(s) 2014

reactions are the reactions of two elements to form a compound. After all, if two elements are treated with each other, they can either react or not.

Type 2: decomposition reactions (BC → B + C). The second type of simple reaction is decomposition. This reaction is also easy to recognize. Typically, only one reactant is given. A type of energy, such as heat or electricity, may also be indicated. The reactant usually decomposes to its elements, to an element and a simpler compound, or to two simpler compounds.

Type 3: substitution reactions (C + AB → AC + B). Elements have varying abilities to be combined. Among the most reactive metals are the alkali metals and the alkaline earth metals. On the opposite end of the scale of reactivity, among the least active metals or the most stable metals are silver and gold, prized for their lack of reactivity. Reactive means the opposite of stable, but means the same as active.

When a free element reacts with a compound of different elements, the free element will replace one of the elements in the compound if the free element is more reactive than the element it replaces. In general, a free metal will replace the metal in the compound, or a free nonmetal will replace the nonmetal in the compound. A new compound and a new free element are produced.

Type 4: double-substitution reactions (AB + CD → CB + AD). Double-substitution or double-replacement reactions, also called double-decomposition reactions or metathesis reactions, involve two ionic compounds, most often in aqueous solution. In this type of reaction, the cations simply swap anions. The reaction proceeds if a solid or a covalent compound is formed from ions in solution. All gases at room temperature are covalent. Some reactions of ionic solids plus ions in solution can also occur. Otherwise, no reaction takes place. Just as with replacement reactions, double-replacement reactions may or may not proceed. They need a driving force. In replacement reactions the driving force is reactivity; here it is insolubility or covalence.

After describing the natural paradigm that we intent to mimic, the general structure of the proposed chemical optimization algorithm is presented in the next section.

Reference

1. R. Chang, General Chemistry (McGraw-Hill, New York, 2004)

Chapter 4
The Proposed Chemical Reaction Algorithm

Abstract This chapter introduces the chemical reaction algorithm; it describes the main characteristics and definitions. In this work, the main objective is to introduce a novel optimization algorithm based in a paradigm inspired by nature, the chemical reactions.

Keywords Chemical reactions • Optimization algorithm • Chemical algorithm

This chapter introduces the chemical reaction algorithm; it describes the main characteristics and definitions.

In this work, the main objective is to introduce a novel optimization algorithm based in a paradigm inspired by nature, the chemical reactions.

A novel set of intensifier/diversifier mechanisms will be introduced in order to allow the population to converge to an optimal solution within a determined search space. Comparisons with other nature inspired optimization techniques will be made in order to prove good performance of the proposed algorithm.

Once the optimization algorithm is developed, it will be applied to a set of benchmark functions; the obtained results will be compared with the results obtained with genetic algorithms and other optimization paradigms.

In the second stage, this chemical reaction algorithm will be applied to optimize the parameters of the membership functions of a Type-1 and Type-2 fuzzy logic controller for the tracking of a unicycle mobile robot. With this work, we pretend to demonstrate that this novel optimization paradigm is able to perform well in these specific topics and will encourage its use in different soft computing approaches.

At first sight, chemical theory and its definitions may seem complex and none or few related to optimization theory, but only the general schema will be considered in proposing the chemical reaction optimization algorithm.

Dittrich et al. [1], described the basic terms to characterize and classify artificial chemistries. Given the statistical and qualitative features of the reaction laws and element/component representation, our proposed algorithm is an *abstraction* of the chemical reaction process and can be described as a *constructive dynamical*

L. Astudillo et al., *Chemical Optimization Algorithm for Fuzzy Controller Design*,
SpringerBriefs in Computational Intelligence, DOI: 10.1007/978-3-319-05245-8_4,
© The Author(s) 2014

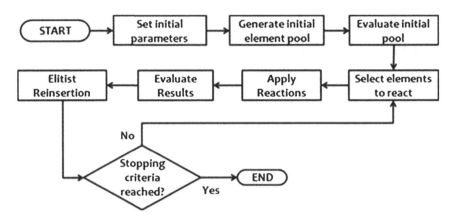

Fig. 4.1 General flowchart of the chemical reaction algorithm

system due the appearance of new components that may change the dynamics of the system. The initial definition of elements and/or compounds depends on the problem itself. These can be represented as binary numbers, integer, floating, etc.

The elements/compounds interact with each other implicitly; the definition of the interaction is independent of the molecular structure and the rules R of interaction in this first attempt will not include real-like parameters, such as temperature, pH, pressure, etc.

As depicted in Chap. 3, atoms are the basic building blocks of everything. When atoms connect with other atoms, they form molecules, which, in combination with other molecules form matter. The key feature here is to observe how these combinations are achieved.

The proposed chemical reaction algorithm is a metaheuristic strategy that performs a stochastic search for optimal solutions within a defined search space. In this optimization strategy, every solution is considered as an element (or compound), and the fitness or performance of the element is represented by the final energy (which is determined according to the objective function). The general flowchart of the algorithm is shown in Fig. 4.1.

The main difference with other optimization techniques [2, 3, 4, 5] is that no external parameters are taken into account to evaluate the results, while other algorithms introduce additional parameters (kinetic/potential energies, mass conservation, thermodynamic characteristics, etc.), this is a very straight forward methodology that takes the characteristics of the chemical reactions (synthesis, decomposition, substitution and double-substitution) to find for optimal solution.

This approach is a static population-based metaheuristic that applies an abstraction of the chemical reactions as intensifiers (substitution, double substitution reactions) and diversifying (synthesis, decomposition reactions) mechanisms. The elitist reinsertion strategy allows the permanence of the best elements and thus the average fitness of the entire element pool increases with every iteration. The algorithm may trigger only one reaction or all of them, depending on the nature of the problem to solve, in example; we may use only the decomposition reaction sub-routine to find the minimum value of a mathematical function.

The basic components of this methodology are explained briefly in the following lines.

At the beginning of the algorithm, a set of elements/compounds is randomly generated under the normal distribution space of possible solutions and this can be represented by:

$$X = \{x_1, x_2, \ldots, x_n\}, \tag{4.1}$$

where x_n represents the element/compound.

The number and representation of the initial elements may vary depending on the complexity of the problem to solve.

A selection procedure is applied to "induce" a reaction between the elements/compounds. The selection process used in the experiments presented in this thesis was the stochastic universal sampling.

Every nature inspired paradigm has their own way to encode candidate solutions. When these parameters are defined, a set of processes or procedures are applied to lead the population to an optimal result. The main components of this chemical reaction algorithm are described below.

4.1 Elements/Compounds

These are the basic components of the algorithm. Each element or compound represents a solution within the search space. The initial definition of elements and/or compounds depends on the problem itself and can be represented as binary numbers, integer, floating, etc. They interact with each other implicitly; this is, the definition of the interaction is independent of the real molecular structure; in this approach the potential and kinetic energies and other molecular characteristics are not taken into account.

4.2 Chemical Reactions

A chemical reaction is a process in which at least one substance changes its composition and its sets of properties, in this approach, the chemical reactions behave as intensifiers (substitution, double substitution reactions) and diversifying (synthesis, decomposition reactions) mechanisms. The four chemical reactions considered in this approach are the *synthesis, decomposition, single and double substitution reactions*. The objective of these operators is exploring or exploiting new possible solutions within a slightly larger hypercube than the original elements/compounds, but within the previously specified range.

The *synthesis* and *decomposition* reactions are used to diversify the resulting solutions; these procedures showed to be highly effective to rapidly lead the results to a desired value. They can be described as follows.

4.3 Synthesis Reactions

Is a reaction of two reactants to produce one product. By combining two (or more) elements, this procedure allows to explore higher valued solutions within the search space. The result can be described as a compound $(B + C \rightarrow BC)$. The pseudocode for the synthesis reaction procedure is as follows:

Synthesis_Procedure
Input: *selected_elements, synthesis_rate*
1. $n = $ size *(selected_elements)*
2. $i = $ floor $(n/2)$
3. **for** $j = 1$ to $i - 1$
4.   Synthesis $= selected\_elements_j + selected\_elements_{j+1}$
5. $j = j + 2$
6. **end for**
Output: *Synthesis_vector*

4.4 Decomposition Reactions

In this reaction, typically, only one reactant is given, it allows a compound to be decomposed into smaller instances $(BC \rightarrow B + C)$. The pseudocode for the decomposition reaction procedure is as follows:

Decomposition_Procedure
Input: *selected_elements, decomposition_rate*
1. $n = $ size *(selected_elements)*
2. Get *randval* randomly in interval [0, 1]
3. **for** $i = 1$ to n
4. $Deco_1 = selected\_elements_i$ x *randval*
5. $Deco_2 = selected\_elements_i$ x $(1 - randval)$
6. $i = i + 1$
7. **end for**
Output: *Decomposition_vector* $(Deco_1, Deco_2)$

The *single* and *double substitution* reactions allow the algorithm to search for optima around a good previously found solution and they're described as follows.

4.5 Single-Substitution Reactions

When a free element reacts with a compound of different elements, the free element will replace one of the elements in the compound if the free element is more reactive than the element it replaces. A new compound and a new free element are produced; during the algorithm, a compound and an element are selected and a decomposition reaction is applied to the compound; two elements are generated from this operation. Then, one of the new generated elements is combined with the non-decomposed selected element ($C + AB \rightarrow AC + B$). The pseudocode for the single-substitution reaction procedure is as follows:

SingleSubstitution_Procedure
Input: *selected_elements, singlesubstitution_rate*
1. $n =$ size (*selected_elements*)
2. $i =$ floor ($n/2$)
3. $a = selected\_elements_1, selected\_elements_2, ..., selected\_elements_i$
4. $b = selected\_elements_{i+1}, selected\_elements_{i+2}, ..., selected\_elements_{ix2}$
5. Apply *Decomposition_Procedure* to a; Get $Deco_1, Deco_2$
6. Apply *Synthesis_Procedure* ($b + Deco_1$); Get *Synthesis_vector*
Output: *SingleSubstitution _vector* (*Synthesis_vector, $Deco_2$*)

4.6 Double-Substitution Reactions

Double-substitution or double-replacement reactions, also called double-decomposition reactions or metathesis reactions involve two ionic compounds, most often in aqueous solution. In this type of reaction, the cations simply swap anions; during the algorithm, a similar process that in the previous reaction happens, the difference is that in this reaction both of the selected compounds are decomposed and the resulting elements are combined between each other ($AB + CD \rightarrow CB + AD$). The pseudocode for the double-substitution reaction procedure is as follows:

DoubleSubstitution_Procedure
Input: *selected_elements, doublesubstitution_rate*
1. $n =$ size (*selected_elements*)
2. $i =$ floor ($n/2$)
3. $a = selected\_elements_1, selected\_elements_2, ..., selected\_elements_i$
4. $b = selected\_elements_{i+1}, selected\_elements_{i+2}, ..., selected\_elements_{ix2}$
5. Apply *Decomposition_Procedure* to a and b; Get ($Deco_1, Deco_2$), ($Deco_1', Deco_2'$)
6. Apply *Synthesis_Procedure* ($Deco_1 + Deco_1'$), ($Deco_2 + Deco_2'$) Get *Synthesis_vector_1, Synthesis_vector_1'*
Output: *SingleSubstitution _vector* (*Synthesis_vector_1, Synthesis_vector_1'*)

Table 4.1 Main elements of several nature inspired paradigms

Paradigm	Parameter representation	Basic operations
GA	Genes	Crossover, mutation
ACO	Ants	Pheromone
PSO	Particles	Cognitive coefficient, social coefficient
GP	Trees	Crossover, mutation (in some cases)
CRM	Elements, compounds	Reactions (combination, decomposition, substitution, double-substitution)

Throughout the execution of the algorithm, whenever a new set of elements/ compounds are created, an elitist reinsertion criteria is applied, allowing the permanence of the best elements and thus the average fitness of the entire element pool increases through iterations.

Analog to some other nature inspired algorithms for example, genetic algorithms, where genes are the basic units and are encoded as strings or chromosomes, in this proposed chemical algorithm, the basic units or candidate solutions are represented by elements or/and compounds, and the metaphor of chemical reactions is used as a procedure to approximate the solution to a desired optima.

In order to have a better picture of the general schema for this proposed chemical reaction algorithm (CRA), a comparison with other nature inspired paradigms is shown in Table 4.1.

References

1. P. Dittrich, J. Ziegler, W. Banzhaf, Artificial Chemistries-a review. Artificial Life **7**, 225–275 (2001)
2. N.-Y. Shi, C.-P. Chu, A molecular solution to the hitting-set problem in DNA-based supercomputing. Inf. Sci. **180**, 1010–1019 (2010)
3. L. Yamamoto, Evaluation of a catalytic search algorithm. in *Proceeding of the 4th International Workshop on Nature Inspired Cooperative Strategies for Optimization, NICSO 2010*, (Granada Spain, 2010), pp. 75–87
4. T. Meyer, L. Yamamoto, W. Banzhaf, C. Tschudin, Elongation control in an algorithmic chemistry, advances in artificial life. Darwin Meets von Neumann, Lecture Notes in Computer Science, vol. 5777 (2011), pp. 273–280
5. J. Xu, A.Y.S. Lam, V.O.K. Li, Chemical reaction optimization for the grid scheduling problem. in *IEE Communication Society, ICC 2010*, (2010), pp. 1–5

Chapter 5
Application Problems

Abstract This chapter introduces the chemical reaction algorithm; it describes the main characteristics and definitions. In this work, the main objective is to introduce a novel optimization algorithm based in a paradigm inspired by nature, the chemical reactions.

Keywords Chemical optimization • Chemical reactions • Function optimization

In this chapter, some applications problems are described. The chemical reaction algorithm has been applied to these diverse applications and the results and observations are presented in the next chapter.

5.1 Complex Benchmark Functions

The quality of the optimization procedures (existing and new) is often evaluated using reference standard problems, such as complex functions.

These complex test functions can be categorized as follows:

a. Unimodal, convex, multidimensional.
b. Multimodal, bi-dimensional with a small number of local minima.
c. Multimodal, bi-dimensional with a large number of local minima.
d. Multimodal, multidimensional, with a large number of local minima.

Table 5.1 shows the complex benchmark functions used in this work to evaluate the performance of the chemical reaction algorithm.

L. Astudillo et al., *Chemical Optimization Algorithm for Fuzzy Controller Design*,
SpringerBriefs in Computational Intelligence, DOI: 10.1007/978-3-319-05245-8_5,
© The Author(s) 2014

Table 5.1 Benchmark test functions

Test function	Definition	Search domain	f_{min}
f_1	$f(x) = \sum\limits_{i=1}^{n} x_i^2$	$[-100, 100]$	0
f_2	$f(x) = \sum\limits_{i=1}^{n} i * x_i^2$	$[-100, 100]$	0
f_3	$f(x) = \sum\limits_{i=1}^{n} \left(\sum\limits_{j=1}^{i} x_j \right)^2$	$[-65, 65]$	0
f_4	$f(x) = \sum\limits_{i=1}^{n} 5i * x_i^2$	$[-100, 100]$	0
f_5	$f(x) = \sum\limits_{i=1}^{n-1} \left[100\left(x_{i+1} - x_i^2\right)^2 + (1 - x_i)^2 \right]$	$[-30, 30]$	0
f_6	$f(x) = 10n + \sum\limits_{i=1}^{n} \left[x_i^2 - 10\cos\left(2\pi x_i\right) \right]$	$[-100, 100]$	0
f_7	$f(x) = \sum\limits_{i=1}^{n} -x_i \sin\left(\sqrt{|x_i|}\right)$	$[-600, 600]$	$-n \cdot 418.9829$
f_8	$f(x) = \sum\limits_{i=1}^{n} \frac{x_i^2}{4000} - \prod\limits_{i=1}^{n} \cos\left(\frac{x_i}{\sqrt{i}}\right) + 1$	$[-600, 600]$	0
f_9	$f(x) = \sum\limits_{i=1}^{n} |x_i|^{(i+1)}$	$[-100, 100]$	0
f_{10}	$f(x) = -a \cdot e^{-b \cdot \sqrt{\frac{\sum\limits_{i=1}^{n} x_i^2}{n}}} - e^{\frac{\sum\limits_{i=1}^{n} \cos\left(c \cdot x_i\right)}{n}} + a + e^1$	$[-100, 100]$	0

5.2 Control of an Autonomous Mobile Robot Using Fuzzy Logic

Mobile robots are nonholonomic systems due to the constraints imposed on their kinematics. The equations describing the constraints cannot be integrated symbolically to obtain explicit relationships between robot positions in local and global coordinate's frames. Hence, control problems that involve them have attracted attention in the control community in the last years.

Different methods have been applied to solve motion control problems. Kanayama et al. [1] propose a stable tracking control method for a nonholonomic vehicle using a Lyapunov function. Lee et al. [2] solved tracking control using backstepping and in [3] with saturation constraints. Furthermore, most reported designs rely on intelligent control approaches such as Fuzzy Logic Control [4–9] and Neural Networks [10, 11].

However the majority of the publications mentioned above, have concentrated on kinematic models of mobile robots, which are controlled by the velocity input, while less attention has been paid to the control problems of nonholonomic dynamic

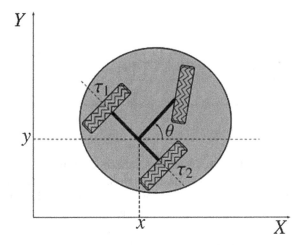

Fig. 5.1 Wheeled mobile robot

systems, where forces and torques are the true inputs: Bloch and Drakunov [12] and Chwa [13], used a sliding mode control to the tracking control problem. Fierro and Lewis [14] propose a dynamical extension that makes possible the integration of kinematics and torque controller for a nonholonomic mobile robot. Fukao et al. [15], introduced an adaptive tracking controller for the dynamic model of mobile robot with unknown parameters using backstepping.

Further publications [16–18] have applied bio-inspired techniques to solve the tracking problem for the dynamic model of a unicycle mobile robot, using a fuzzy logic controller that provides the required torques to reach the desired velocity and trajectory inputs (Fig. 5.1).

5.2.1 Definition of the Mobile Robot

The model considered is that of a unicycle mobile robot (see Fig. 5.2) that has two driving wheels fixed to the axis and one passive orientable wheel that is placed in front of the axis and normal to it [19].

The two fixed wheels are controlled independently by the motors, and the passive wheel prevents the robot from overturning when moving on a plane.

It is assumed that the motion of the passive wheel can be ignored from the dynamics of the mobile robot, which is represented by the following set of equations [14]:

$$\dot{q} = \begin{vmatrix} \cos\theta & 0 \\ \sin\theta & 0 \\ 0 & 1 \end{vmatrix} \begin{vmatrix} v \\ w \end{vmatrix} \tag{5.1}$$

$$M(q)\dot{v} + V(q,\dot{q})v + G(q) = \tau \tag{5.2}$$

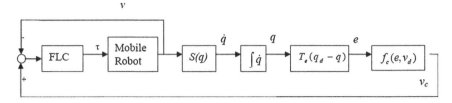

Fig. 5.2 Tracking control structure

where $q = [x, y, \theta]^T$ is the vector of generalized coordinates which describes the robot position, *(x, y)* are the Cartesian coordinates, which denote the mobile center of mass and θ is the angle between the heading direction and the *x*-axis (which is taken counterclockwise form); $v = [v, w]^T$ is the vector of velocities, *v* and *w* are the linear and angular velocities respectively; $\tau \in R^r$ is the input vector, $M(q) \in R^{nxn}$ is a symmetric and positive-definite inertia matrix, $V(q, \dot{q}) \in R^{nxn}$ is the centripetal and Coriolis matrix, $G(q) \in R^n$ is the gravitational vector. Equation (5.1) represents the kinematics or steering system of a mobile robot.

Notice that the no-slip condition imposed a nonholonomic constraint described by (5.3), that it means that the mobile robot can only move in the direction normal to the axis of the driving wheels.

$$\dot{y}\cos\theta - \dot{x}\sin\theta = 0 \tag{5.3}$$

5.2.2 Tracking Controller of Mobile Robot

The control objective will be established as follows: Given a desired trajectory $q_d(t)$ and the orientation of the mobile robot we must design a controller that applies an adequate torque τ such that the measured positions $q(t)$ achieve the desired reference $q_d(t)$ represented as (5.4):

$$\lim_{t \to \infty} \|q_d(t) - q(t)\| = 0 \tag{5.4}$$

To reach the control objective, we are based on the procedure of [14], we are deriving a $\tau(t)$ of a specific $v_c(t)$ that controls the steering system (5.1) using a Fuzzy Logic Controller (FLC). A general structure of tracking control system is presented in the Fig. 5.3.

5.2.3 Control of the Kinematic model

We are based on the procedure proposed by Kanayama et al. [1] and Nelson et al. [20] to solve the tracking problem for the kinematic model $v_c(t)$. Suppose that the desired trajectory q_d satisfies (5.5):

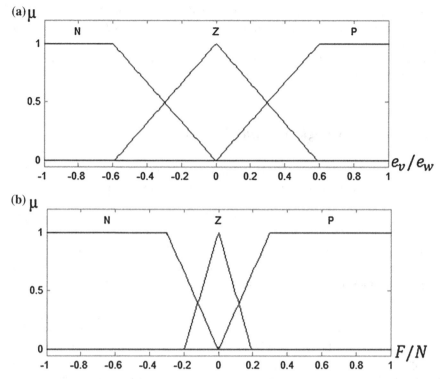

Fig. 5.3 Membership functions of the **a** input e_v and e_w, and **b** output variables F and N

$$\dot{q}_d = \begin{vmatrix} \cos\theta_d & 0 \\ \sin\theta_d & 0 \\ 0 & 1 \end{vmatrix} \begin{vmatrix} v_d \\ w_d \end{vmatrix} \tag{5.5}$$

Using the robot local frame (the moving coordinate system x–y in Fig. 5.2), the error coordinates can be defined as (5.6):

$$e = T_e(q_d - q), \begin{vmatrix} e_x \\ e_y \\ e_\theta \end{vmatrix} = \begin{vmatrix} \cos\theta & \sin\theta & 0 \\ -\sin\theta & \cos\theta & 0 \\ 0 & 0 & 1 \end{vmatrix} \begin{vmatrix} x_d - x \\ y_d - y \\ \theta_d - \theta \end{vmatrix} \tag{5.6}$$

And the auxiliary velocity control input that achieves tracking for (5.1) is given by (5.7):

$$v_c = f_c(e, v_d), \begin{vmatrix} v_c \\ w_c \end{vmatrix} = \begin{vmatrix} v_d + \cos e_\theta + k_1 e_x \\ w_d + v_d k_2 e_y + v_d k_3 \sin e_\theta \end{vmatrix} \tag{5.7}$$

where k_1, k_2 and k_3 are positive gain constants.

The first approach in the optimization for the fuzzy controller of the mobile robot is to apply our proposed method to obtain the values of k_i ($i = 1, 2, 3$) for optimal behavior of the controller.

Once we have found these gain constraints, the second step is to find the values of the membership functions of the fuzzy controller.

5.2.4 The Fuzzy Logic Tracking Controller

The purpose of the FLC is to find a control input τ such that the current velocity vector v is able to reach the velocity vector v_c this is denoted as (5.8):

$$\lim_{t \to \infty} \|v_c - v\| = 0 \qquad (5.8)$$

The inputs variables of the FLC correspond to the velocity errors obtained of (5.6) (denoted as ev and ew: linear and angular velocity errors respectively), and 2 outputs variables, the driving and rotational input torques τ (denoted by F and N respectively).

The initial membership functions (MF) are defined by 1 triangular and 2 trapezoidal functions for each variable involved. In future work the shape of the membership functions will be selected by the algorithm as part of the optimization.

Figure 5.3a and b depicts the MFs in which N, Z, P represent the fuzzy sets (Negative, Zero and Positive respectively) associated to each input and output variable. Table 5.2 shows the upper and lower limits of the used membership functions.

The rule set of the FLC contain 9 rules, which govern the input–output relationship of the FLC and this adopts the Mamdani-style inference engine. We use the center of gravity method to realize defuzzification procedure. In Table 5.3, we present the rule set whose format is established as follows:

Rule i : If ev is G1 and ew is G2 then F is G3 and N is G4   (5.9)

where G1..G4 are the fuzzy sets associated to each variable and $i = 1 \dots 9$.

5.3 Control of an Autonomous Mobile Robot Using Type-2 Fuzzy Logic

The tracking controller obtained by means of fuzzy logic was considered as a base to design a type-2 fuzzy logic controller. The membership function types and parameters of the primary membership functions are the same that resulted in the type-1 fuzzy controller. The parameters that the chemical reaction paradigm will attempt to find are those for the secondary membership functions. Once these parameters are found, the objective is to test the performance of the fuzzy logic controllers (type-1 and type-2) by applying a disturbance to the tracking controller system, and it is given by:

Table 5.2 Upper and lower limits of the membership functions

Membership function	Point	Lower limit	Upper limit
Trapezoidal	a	-1000	-1000
	b	-1000	-1000
	c	-800	-300
	d	-300	250
Triangular	a	-800	-200
	b	-50	50
	c	200	800
Trapezoidal	a	-250	300
	b	300	800
	c	1000	1000
	d	1000	1000

Table 5.3 Fuzzy rule set

e_v/e_w	N	C	P
N	N/N	N/Z	N/P
Z	Z/N	Z/Z	Z/P
P	P/N	P/Z	P/P

$$\text{Perturbation} = \varepsilon \sin \omega t \qquad (5.10)$$

where $t = $ time, in an interval of 1–10 s; ε varies from 0.05 to 41, and $\omega = 1$.

The results of the proposed chemical reaction algorithm applied to the benchmark equations and the fuzzy logic Type-1 and Type-2 controllers are shown in Chap. 6.

References

1. Y. Kanayama, Y. Kimura, F. Miyazaki, T. Noguchi, A stable tracking control method for a non-holonomic mobile robot. in *Proceedings of the IEEE/RSJ International Workshop on Intelligent Robots and Systems.*(Osaka, Japan, 1991), pp 1236–1241
2. T.-C. Lee, C.H. Lee, C.-C. Teng, Tracking control of mobile robots using the backsteping technique. in *Proceedings of the 5th International Conference Control, Automation, Robotics Vision.* (Singapore, Dec 1998), pp 1715–1719
3. T.-C. Lee, K. Tai, Tracking Control of Unicycle-Modeled Mobile robots Using a Saturation Feedback Controller. IEEE Trans. Control Syst. Technol. 9(2), 305–318 (2001)
4. S. Bentalba, A. El Hajjaji, A. Rachid, Fuzzy control of a mobile robot: a new approach. in *Proceedings of the IEEE International Conference on Control Applications*, (Hartford, CT, Oct 1997), pp 69–72
5. S. Ishikawa, A method of indoor mobile robot navigation by fuzzy control. in *Proceedings of the International Conference Intelligent Robotic System.* (Osaka, Japan, 1991), pp. 1013–1018
6. T.H. Lee, F.H.F. Leung, P.K.S. Tam, Position control for wheeled mobile robot using a fuzzy controller. IEEE 2, 525–528 (1999)
7. S. Pawlowski, P. Dutkiewicz, K. Kozlowski, W. Wroblewski, Fuzzy logic implementation in mobile robot control. in *2nd Workshop on Robot Motion and Control*, (Oct 2001), pp 65–70

8. C.-C. Tsai, H.-H. Lin, C.-C. Lin, Trajectory tracking control of a laser-guided wheeled mobile robot, in *Proceedings of the IEEE International Conference on Control Applications*. (Taipei, Taiwan, Sept 2004), pp 1055–1059

9. S.V. Ulyanov, S. Watanabe, V.S. Ulyanov, K. Yamafuji, L.V. Litvintseva, G.G. Rizzotto, Soft computing for the intelligent robust control of a robotic unicycle with a new physical measure for mechanical controllability, soft computing, vol. 2. (Springer, 1998), pp 73–88

10. R. Fierro, F.L. Lewis, Control of a non holonomic mobile robot using neural networks. IEEE Trans. Neural Networks **9**(4), 589–600 (1998)

11. K.T. Song, L.H. Sheen, Heuristic fuzzy-neural network and its application to reactive navigation of a mobile robot. Fuzzy Sets Syst. **110**(3), 331–340 (2000)

12. A.M. Bloch, S. Drakunov, Tracking in non holonomic dynamic system via sliding modes. in *Proceedings IEEE Conference on Decision and Control*, (Brighton, UK, 1991), pp. 1127–1132

13. D. Chwa, Sliding-mode tracking control of non holonomic wheeled mobile robots in polar coordinates. IEEE Trans. Control Syst. Tech. **12**(4), 633–644 (2004)

14. R. Fierro, F.L. Lewis, Control of a non holonomic mobile robot: backstepping kinematics into dynamics. in *Proceedings of the 34th Conference on Decision and Control*, (New Orleans, LA, 1995)

15. T. Fukao, H. Nakagawa, N. Adachi, Adaptive tracking control of a nonnholonomic mobile robot. IEEE Trans. Robot. Autom. **16**(5), 609–615 (2000)

16. L. Astudillo, O. Castillo, L. Aguilar, Intelligent control for a perturbed autonomous wheeled mobile robot: a type-2 fuzzy logic approach. Nonlinear Studies **14–1**, 37–48 (2007)

17. R. Martinez, O. Castillo, L. Aguilar, Optimization of type-2 fuzzy logic controllers for a perturbed autonomous wheeled mobile robot using genetic algorithms. Inf. Sci. **179–13**, 2158–2174 (2009)

18. O. Castillo, R. Martinez-Marroquin, P. Melin, J. Soria, Comparative study of bio-inspired algorithms applied to the optimization of type-1 and type-2 fuzzy controllers for an autonomous mobile robot. Bio-inspired Hybrid Intelligent Systems for Image Analysis and Pattern Recognition, Studies in Computational Intelligence **256**(2009), 247–262 (2009)

19. G. Campion, G. Bastin, B. D'Andrea-Novel, Structural properties and classification of kinematic and dynamic models of wheeled mobile robots. IEEE Trans. Robot. Autom. **12**(1), 47–62 (1996)

20. W. Nelson, I. Cox, Local path control for an autonomous vehicle, in *Proceedings of the IEEE Conference on Robotics and Automation*, (1988), pp. 1504–1510

Chapter 6
Simulation Results

Abstract This chapter shows the simulation results obtained with the chemical optimization algorithm for the optimization of benchmark functions and robot fuzzy control design.

Keywords Simulation results • Function optimization • Fuzzy control design • Robot control

This chapter shows the results obtained for the application problems described in Chap. 4. The results are presented in the same order that the application problems were introduced.

6.1 Results of the CRA Applied to the Complex Benchmark Functions

Making a fair comparison between two or more optimization algorithms can be hard due to several reasons; one of them is the initial set of parameters, which plays an important role in the performance of the algorithm; the number of iterations and their overall characteristics that make the algorithms different from each other (e.g. selection pressure, replacement criteria, update of certain parameters, domain exploration/exploitation strategies, etc.).

In order to validate the proposed algorithm, we implemented it in Matlab. Simulations were made; Table 6.1 shows the initial parameters of each set, which consist in 50 runs of the algorithm for every function. The selection method was the stochastic universal sampling and an elitist reinsertion criteria was followed, based on the fitness of each resultant element/compound.

As seen on Table 6.1, the chemical reaction rates didn't change with every set, this was considered to evaluate the behavior of the algorithm, these rates determine the quantity of elements to be selected to react, e.g. for a pool of 50 elements, a rate

Table 6.1 Initial set of parameters of the compared optimization algorithms and the (CRA)

Parameter	Set 1	Set 2	Set 3
Population size	10	10	100
Maximum iteration	10	100	10
Synthesis rate	0.2	0.2	0.2
Decomposition rate	0.2	0.2	0.2
Single substitution rate	0.2	0.2	0.2
Double substitution rate	0.2	0.2	0.2

Table 6.2 Some initial set of parameters of the compared optimization algorithms and the chemical reaction algorithm (CRA)

Parameter	GA	PSO	GSA
Population size (range)	20–150	6–40	50
Maximum iteration (range)	80–200	Not mentioned	1000
Selection method	RWS	Not mentioned	RWS
Mutation	0.8	–	–
Crossover	0.5	–	–
Cognitive acceleration	–	1	–
Social acceleration	–	0.5	–
Velocity at the beginning	–	0.95	–
Constriction factor	–	1	–
G_0 (Gravitational constant), α	–	–	100, 20

of 0.2 will let 10 elements to be selected for reaction. In some cases, a population of 10 elements was enough to get good results, on some other cases and according on the test function's complexity, a larger population was needed.

In this research work, we compared the results obtained with the chemical reaction algorithm and an improved evolutionary method with fuzzy logic for combining PSO and GA as well as a gravitational search algorithm (GSA). All algorithms have their specific parameter representation. In Table 6.2; the values for the initial parameters on the compared algorithms are presented.

The simulations were performed 30 times each. Tables A.1, A.2 and A.3 in the Annex A section show the complete results, including the best and worst solutions found by the algorithm, as well as the mean and standard deviation for the test functions of Table 5.1 showed in the previous chapter.

In Tables 6.3 and 6.4, the results obtained with the Chemical Reaction algorithm applying the parameters of set 2 and 256 dimensions compared with an improved evolutionary method with fuzzy logic for combining PSO and GA [1], and a Gravitational Search Algorithm [2] for functions f_1 to f_5 and f_6 to f_{10} respectively are presented.

Some authors prefer not to include some specific values like the worst result obtained, the standard deviation, and others, as well as the results for some test functions were omitted; in this thesis we included these values in order to provide a start point for comparisons against other optimization strategies.

Table 6.3 Comparison of results between the chemical reaction algorithm (CRA), PSO+GA and GSA, applied to the set to the benchmark functions f_1 to f_5 of Table 5.1

Function	Algorithm	Dim	Best	Worst	Mean	Std. Dev.
f_1	FPSO+FGA	256	0.9601	2.996	0.6060	N/A
	GSA	30	7.3E-11	N/A	2.1E-10	N/A
	CRA-Set 1	256	1.09E-04	1.03E+03	2.01E+02	2.43E+02
	CRA-Set 2	**256**	**4.75E-35**	**5.96E-28**	**2.68E-29**	**9.09E-29**
	CRA-Set 3	256	1.17E-01	3.83E+02	7.12E+01	8.57E+01
f_2	FPSO+FGA	256	N/A	N/A	N/A	N/A
	GSA	30	N/A	N/A	N/A	N/A
	CRA-Set 1	256	1.26	4.69E+04	6.55E+03	1.01E+04
	CRA-Set 2	**256**	**4.58E-34**	**2.29E-25**	**9.17E-27**	**3.59E-26**
	CRA-Set 3	256	8.69E-01	1.05E+06	1.59E+05	2.47E+05
f_3	FPSO+FGA	256	6.9500	97.09	8.6008	N/A
	GSA	30	0.16E+03	N/A	0.16E+03	N/A
	CRA-Set 1	256	2.03E-02	1.46E+04	3.83E+03	4.04E+03
	CRA-Set 2	**256**	**1.36E-32**	**2.80E-25**	**6.04E-27**	**3.96E-26**
	CRA-Set 3	256	2.85E+02	7.47E+05	1.27E+05	1.74E+05
f_4	FPSO+FGA	256	N/A	N/A	N/A	N/A
	GSA	30	N/A	N/A	N/A	N/A
	CRA-Set 1	256	2.00E+01	3.91E+05	4.37E+04	7.02E+04
	CRA-Set 2	**256**	**1.66E-31**	**1.09E-24**	**2.38E-26**	**1.54E-25**
	CRA-Set 3	256	2.02E+02	1.13E+07	1.55E+06	2.40E+06
f_5	**FPSO+FGA**	**256**	**3.0906**	**9.0456**	**6.7687**	**N/A**
	GSA	30	25.16	N/A	25.16	N/A
	CRA-Set 1	256	2.55E+02	4.44E+03	4.91E+02	6.41E+02
	CRA-Set 2	256	2.55E+02	2.55E+02	2.55E+02	1.50E-02
	CRA-Set 3	256	2.55E+02	5.07E+02	2.89E+02	5.94E+01

Table 6.4 Comparison of results between the chemical reaction algorithm (CRA), PSO+GA and GSA, applied to the set to the benchmark functions f_6 to f_{10} of Table 5.1

Function	Algorithm	Dim	Best	Worst	Mean	Std. Dev.
f_6	FPSO+FGA	256	1.9878	11.800	4.0334	N/A
	GSA	30	15.32	N/A	15.32	N/A
	CRA-Set 1	256	7.80	3.45E+03	1.45E+03	7.61E+02
	CRA-Set 2	**256**	**0.00**	**0.00**	**0.00**	**0.00**
	CRA-Set 3	256	4.23	2.10E+03	8.87E+02	6.02E+02
f_7	**FPSO+FGA**	**256**	**0.7096**	**2.9970**	**1.0996**	**N/A**
	GSA	30	−2.8E+03	N/A	−1.1E+03	N/A
	CRA-Set 1	256	−1.57E+04	−4.13E+03	−8.57E+03	2.79E+03
	CRA-Set 2	256	−7.23E+05	−9.54E+03	−1.19E+05	1.53E+05
	CRA-Set 3	256	−1.93E+04	−8.46E+03	−1.23E+04	2.04E+03
f_8	FPSO+FGA	256	2.6779	12.999	4.5001	N/A
	GSA	30	0.29	N/A	0.29	N/A
	CRA-Set 1	256	1.50E-01	1.00E+01	2.79	2.49
	CRA-Set 2	**256**	**0.00**	**0.00**	**0.00**	**0.00**
	CRA-Set 3	256	3.80E-06	5.16	1.31	1.22

(continued)

Table 6.4 (continued)

Function	Algorithm	Dim	Best	Worst	Mean	Std. Dev.
f_9	FPSO+FGA	256	N/A	N/A	N/A	N/A
	GSA	30	N/A	N/A	N/A	N/A
	CRA-Set 1	256	8.40E-11	4.59E-06	5.14E-07	1.05E-06
	CRA-Set 2	**256**	**8.55E-61**	**1.70E-50**	**3.50E-52**	**2.41E-51**
	CRA-Set 3	256	9.78E-13	1.82E-07	1.25E-08	2.86E-08
f_{10}	FPSO+FGA	256	0.0666	0.9999	0.560	N/A
	GSA	30	6.9E-06	N/A	1.1E-05	N/A
	CRA-Set 1	256	7.20E-02	1.34E+01	4.14	3.18
	CRA-Set 2	**256**	**1.35E-05**	**6.48E-01**	**5.70E-02**	**1.31E-01**
	CRA-Set 3	256	1.87E-03	2.04	2.66E-01	4.23E-01

From Tables 6.3, 6.4 and Tables A.1 to A.3 in the annex A section, it can be observed that the proposed chemical reaction algorithm was able to reach values very close to the global minimum compared with other optimization approaches.

In the case of f_5 and f_7, the algorithm was not able to obtain a smaller value; in which the improved evolutionary method with fuzzy logic for combining FPSO and GA performed better.

By the results showed in Table A.2 in the annex A section, it can be observed that when increasing the number of elements evaluated, the tendencies are that the proposed algorithm is capable of finding smaller values even with a small set of iterations (10 in this case).

Figures 6.1, 6.2, 6.3, 6.4, 6.5 shows the convergence graphics of the chemical reaction algorithm applying the parameters of set 2 and 256 dimensions applied to the test functions mentioned in Table 5.1.

6.2 Results of the CRA Applied to the Fuzzy Control of an Autonomous Mobile Robot

6.2.1 Finding k_1, k_2, k_3

Several tests of the chemical optimization paradigm were made to test the performance of the tracking controller. First, we need to find the values of k_i ($i = 1, 2, 3$) showed in Eq. (5.7), which shall guarantee convergence of the error e to zero.

To evaluate the constants obtained by the algorithm, the mobile robot tracking system, which consists in Eqs. (5.6) and (5.7) was modeled using Simulink®. Figure 6.2 shows the closed loop for the tracking controller (Fig. 6.6).

The conditions of the simulation are given in the following equations.

The positive-definite inertia matrix is given in Eq. (6.1).

$$M(q) = \begin{bmatrix} 0.3749 & -0.0202 \\ -0.0202 & 0.3749 \end{bmatrix} \tag{6.1}$$

Fig. 6.1 **a** f_1; **b** f_2, convergence of the chemical algorithm applying the parameters of set 2 and 256 dimensions applied to the test functions mentioned in Table 6.2

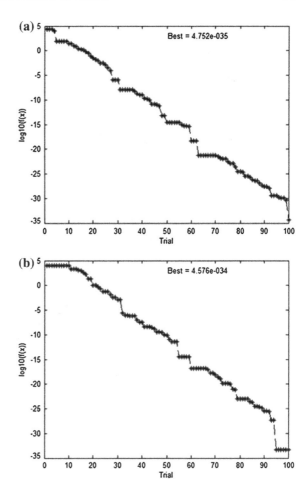

Equation (6.2) shows the centripetal and coriolis forces.

$$C(q, \dot{q}) = \begin{bmatrix} 0 & 0.1350\dot{\theta} \\ 0.1350\dot{\theta} & 0 \end{bmatrix} \qquad (6.2)$$

Equation (6.3) shows a diagonal positive-definite matrix.

$$D = \begin{bmatrix} 10 & 0 \\ 0 & 10 \end{bmatrix} \qquad (6.3)$$

Equation (6.4) shows the desired trajectory.

$$v_d(t) = \left\{ \begin{array}{c} v_d(t) = 0.25 - 0.25 \, Cos\left(\frac{2\pi t}{5}\right) \\ w_d(t) = 0 \end{array} \right\} \qquad (6.4)$$

Fig. 6.2 a f_3; b f_4, convergence of the chemical algorithm applying the parameters of set 2 and 256 dimensions applied to the test functions mentioned in Table 6.2

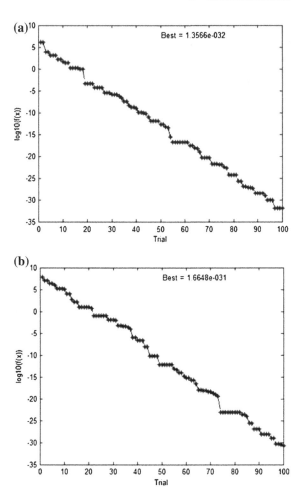

The conditions to evaluate each result, which correspond to the final position error, are given by Eq. (6.5).

$$EP = \sum_{i=1}^{n} \frac{e_x(i) + e_y(i) + e_\theta(i)}{n} \qquad (6.5)$$

For the first set of experiments only the decomposition reaction mechanism was triggered and the decomposition factor was varied; this factor is the quantity of resulting elements after applying a decomposition reaction to a determined "compound"; the only restriction here is that let x be the selected compound and $x'_i (i = 1, 2,...,n)$, the resulting elements; the sum of all values found in the decomposition must be equal to the value of the original compound. This is shown in Eq. (6.6).

$$\sum_{i=1}^{n} x'_i = x \qquad (6.6)$$

Fig. 6.3 **a** f_5; **b** f_6, convergence of the chemical algorithm applying the parameters of set 2 and 256 dimensions applied to the test functions mentioned in Table 6.2

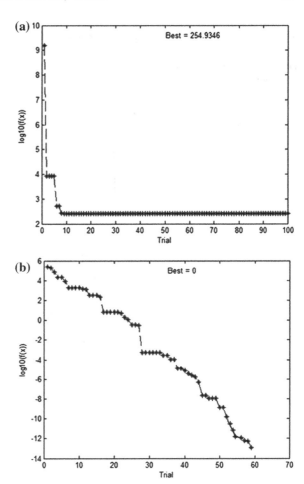

Each experiment was executed 35 times and the test parameters for each set of experiments can be observed in Table 6.5.

The decomposition rate (Dec. Rate) represents the percentage of the pool to be candidate for the decomposition, and the decomposition factor (Dec. Factor) is the number of elements to be decomposed into. In example, for a pool containing 5 initial compounds if the decomposition rate is 0.4, the selected elements to decompose will be 2; then, by applying a decomposition factor of 3, this leads to 6 new elements generated by the decomposition reaction procedure (per every selected element there will be 3 new elements generated). These new elements are reinserted in the original element/compound pool, by applying this criterion, the initial pool of elements increases with every iteration; this is why the initial element pool was set to 10 elements as maximum.

The selection strategy applied was the stochastic universal sampling, which uses a single random value to sample all of the solutions by choosing them at evenly spaced intervals.

Fig. 6.4 **a** f_7; **b** f_8, convergence of the chemical algorithm applying the parameters of set 2 and 256 dimensions applied to the test functions mentioned in Table 6.2

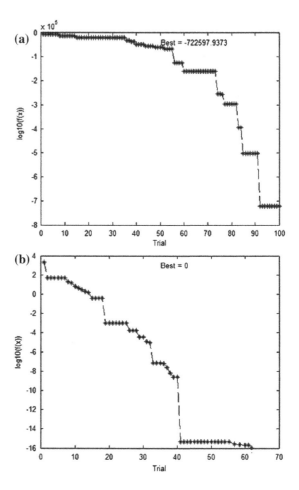

Table 6.6 shows the results after applying the chemical optimization paradigm.

As is observed in Table 6.6, experiment number 6 seems to be the best result because it reached the smaller final error among all experiments. Figure 6.7 shows the final position errors in x, y and *theta* for experiment No. 6 of the first approach.

By analyzing the graphical results of several set of exercises, we noticed that the control obtained for some of them was "smoother" despite the average error value. This was the case for experiment No. 3, in which the final error value was significantly higher than the obtained in experiment No. 6. Figure 6.8 shows the final position errors in x, y and *theta* for experiment No. 3 of the first approach.

Making a comparison between both graphics, we can observe that the average error obtained for *theta* is 0.0338 for experiment No. 6 and 0.0315 for experiment No. 3.

This smoother control of the tracking system could make a big difference in the complete dynamic system of the mobile robot. For this reason these gain constants will be considered for test when searching for the FLC.

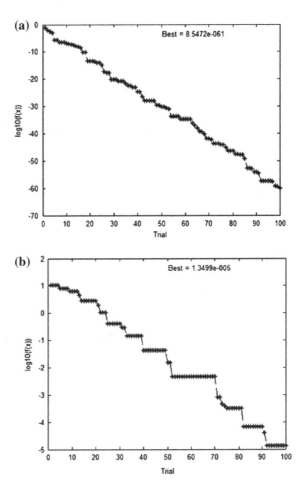

Fig. 6.5 a f_9; **b** f_{10}, convergence of the chemical algorithm applying the parameters of set 2 and 256 dimensions applied to the test functions mentioned in Table 6.2

In order to have a better perspective of the algorithm, the four chemical reaction mechanisms were applied and 10 sets of experiments were made varying the reaction rate values.

In these set of experiments, the decomposition factor was fixed to 2, the size of the pool was maintained to 10 elements/compounds, 50 iterations were performed and the same elitist reinsertion strategy was applied, keeping the compounds/elements with best performance through all the iterations, unless new elements/compounds with better performance are generated.

The test parameters for each set of experiments can be observed in Table 6.7.

The result of varying the parameters for the reaction rates are shown in Table 6.8.

As is observed in Table 6.8, the experiment number 2 obtained the smaller final error. Figures 6.9 and 6.10 show the convergence and final position errors in x, y and *theta* for experiment No. 2 respectively.

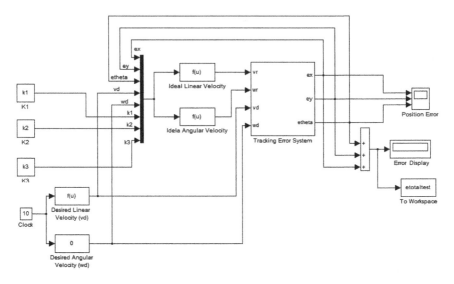

Fig. 6.6   Closed loop for the tracking controller system

	No.	Elements	Iterations	Dec. Factor	Dec. Rate
Table 6.5 Parameters of the chemical reaction optimization	1	2	10	2	0.3
	2	5	10	3	0.3
	3	2	10	2	0.4
	4	2	10	3	0.4
	5	5	10	2	0.4
	6	5	10	3	0.4
	7	5	10	2	0.5
	8	10	10	2	0.5

	No.	Best Error	Mean	k_1	k_2	k_3
Table 6.6 Parameters of the chemical reaction optimization	1	0.0086	1.1568	519	46	8
	2	4.79e-04	0.1291	205	31	31
	3	0.0025	0.5809	36	328	88
	4	0.0012	0.5589	2	206	0
	5	0.0035	0.0480	185	29	5
	6	8.13e-005	0.0299	270	53	15
	7	0.0066	0.1440	29	15	0
	8	0.0019	0.1625	51	3	0

In previous work [3], the gain constant values were found by means of genetic algorithms.

Figure 6.11 shows the shows the behavior of the best overall result applying GAs as optimization method.

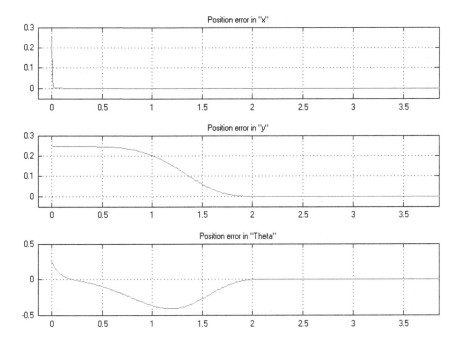

Fig. 6.7 Final position errors in *x*, *y* and *theta* for experiment No. 6 of the first approach

In Table 6.9 we have a comparison of the best results obtained with both algorithms; we can observe that the result with the chemical optimization outperforms the GA in finding the best gain values.

6.2.2 Optimization of a Fuzzy Logic Controller

Once we have found optimal values for the gain constants, the next step is to find the optimal values for the input/output membership functions of the fuzzy controller. Our goal is that in the simulations, the linear and angular velocities reach zero and the mobile robot follow the reference trajectory represented in Eq. (6.4) and showed in Fig. 6.12.

There were 3 experiments with good results and based on those we will apply the CRA to a FLC Type-1. Table 6.10 shows the parameters used in the simulations using the gain constants found by the algorithm.

We will call "Cases" to the group of simulations realized under the same parameters. Case 1 corresponds to the gain constants found by the CRA applying only the decomposition reaction in experiment No. 6; these gains obtained the smallest velocity error (8.13e-005).

Case 2 corresponds to the gain constants that obtained a smaller error for *theta* in experiment No. 3 (0.0315) but not a small velocity error (0.0025796).

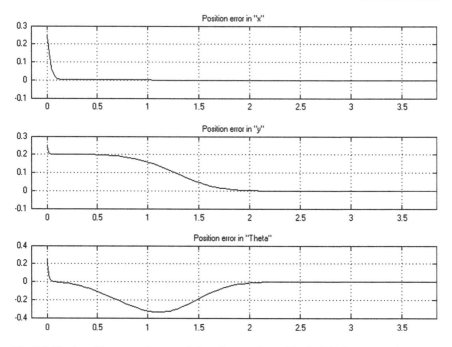

Fig. 6.8   Final position errors in *x*, *y* and *theta* for experiment No. 3 of the first approach

Table 6.7   Parameters of the chemical reaction optimization for the second set of experiments, where the four chemical reactions were applied

Parameters	Synthesis rate	Decomposition rate	Substitution rate	Double substitution rate
Set 1	0.2	0.8	0.2	0.2
Set 2	0.3	0.8	0.6	0.6
Set 3	0.1	0.4	0.1	0.1
Set 4	0.2	0.2	0.5	0.5
Set 5	0.5	0.2	0.6	0.6
Set 6	0.5	0.1	0.8	0.8
Set 7	0.2	0.8	0.5	0.5
Set 8	0.5	0.5	0.5	0.5
Set 9	0.3	0.8	0.8	0.8
Set 10	0.4	0.6	0.7	0.7

Table 6.8   Experimental results of the proposed method for optimizing the values of the gains k_1, k_2, k_3

No.	Best error	Worst error	K_1	K_2	K_3
Set 1	1.83E-07	2.97E-05	382	295	99
Set 2	1.3092E-08	5.15E-06	117	226	132
Set 3	6.69E-07	3.16E-05	113	225	54
Set 4	1.42E-05	1.54E-05	87	272	65
Set 5	1.83E-07	1.32E-06	658	658	236
Set 6	8.88E-06	2.53E-05	28	156	22
Set 7	7.93E-06	4.78E-05	29	134	16
Set 8	1.68E-06	1.58E-05	93	208	45
Set 9	1.02E-05	2.57E-05	161	494	145
Set 10	1.28E-05	2.01E-05	71	285	64

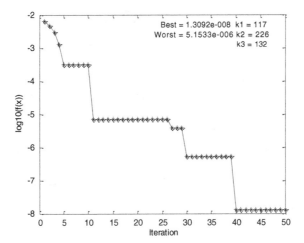

Fig. 6.9   Convergence of experiment No. 2

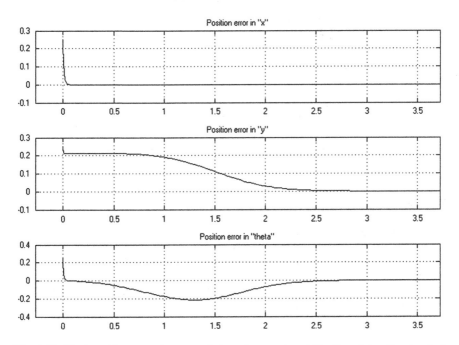

Fig. 6.10 Final position errors in x, y and *theta* for experiment No. 2 applying the chemical reaction algorithm

Case 3 corresponds to the gain constants found by the CRA applying all the decomposition reactions and whose parameters found the smaller velocity error (1.3092E-08). To evaluate each case, we use the average error given by Eq. (6.1).

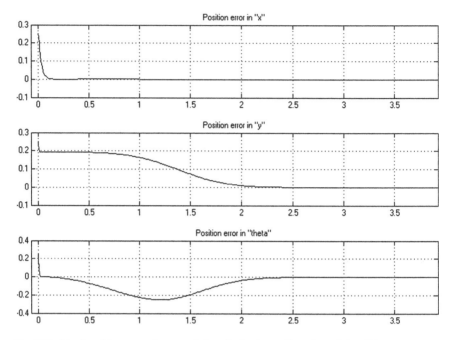

Fig. 6.11 Final position errors in *x*, *y* and *theta* for best overall result applying GAs

Table 6.9 Comparison of best results obtained with GA's and CRA

Parameters	GA	CRA (1st Approach)		CRA (2nd Approach)
Individuals	5	2	2	10
Iterations	15	10	10	50
Crossover rate	0.8	N/A	N/A	N/A
Mutation rate	0.1	N/A	N/A	N/A
Synthesis rate	N/A	Not used	Not used	0.3
Decomposition rate	N/A	0.4	0.4	0.8
Substitution rate		Not used	Not used	0.6
Double-substitution rate	N/A	Not used	Not used	0.6
Decomposition factor	N/A	3	3	2
k1, k2, k3	43, 493, 19	270, 53, 15	36, 328, 88	117, 226, 132
Final error	0.006734	8.13e-005	0.0025796	1.3092E-08

6.2.2.1 Case 1

Table 6.10 shows the parameters used for the simulations of case 1.

For Case 1, the minimum error value found by the chemical reaction algorithm was 0.89855. Figure 6.13 shows the resulted input membership functions found by the proposed optimization algorithm for the linear and angular velocity errors.

Fig. 6.12 Reference trajectory represented by Eq. (6.4)

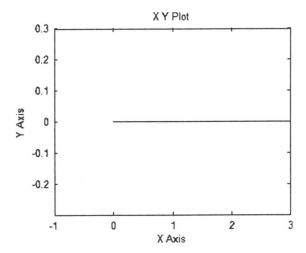

Table 6.10 Simulation parameters and final error of the CRA for Case 1

Parameters	Case 1
Individuals	10
Iterations	15
Synthesis rate	0.2
Decomposition rate	0.4
Substitution rate	0.2
Double-substitution rate	0.2
Decomposition factor	2
k_1, k_2, k_3	270, 53, 15
Final error	0.89855

Figure 6.14 shows the resulted output membership functions found by the proposed optimization algorithm for the left and right torques.

Table 6.11 shows the values of the input/output membership functions for the Type-1 FLC found by the chemical reaction algorithm.

Figure 6.15 shows the obtained trajectory of the best result for Case 1. We can observe that there is no control, the robot is not capable of reach the desired position.

6.2.2.2 Case 2

Table 6.12 shows the parameters used for the simulations of case 2.

For Case 2, the final error found was 0.077178. Figure 6.16 the resulted input membership functions found by the proposed optimization algorithm for the linear and angular velocity errors.

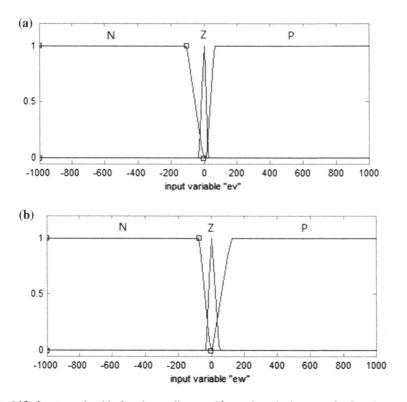

Fig. 6.13 Input membership functions: **a** linear and **b** angular velocity errors for Case 1

Figure 6.17 shows the resulted output membership functions found by the proposed optimization algorithm for the left and right torques.

Table 6.13 shows the values of the input/output membership functions for the Type-1 FLC found by the chemical reaction algorithm.

Figure 6.18 shows the obtained trajectory of the best result for Case 2. We can observe that for this case, the control is sufficient for the mobile robot to reach de desired trajectory.

6.2.2.3 Case 3

Table 6.14 shows the parameters used for the simulations of case 3.

For Case 3, the final error found was 0.061619. Figure 6.19 shows the resulting input membership functions found by the proposed optimization algorithm for the linear and angular velocity errors.

Figure 6.20 shows the resulted output membership functions found by the proposed optimization algorithm for the left and right torques.

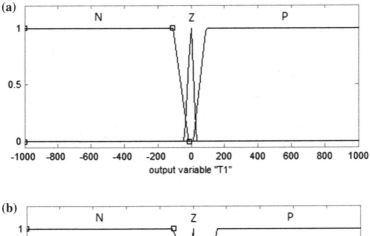

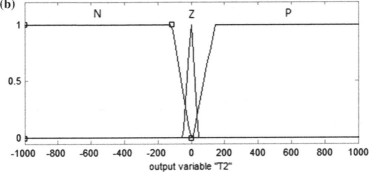

Fig. 6.14 Resulting output membership functions: **a** *left* and **b** *right* torques for Case 1

Table 6.11 Parameters of the membership functions for Type-1 FLC for Case 1	Variable	Membership function
	Linear velocity error e_v	Trapezoidal $[-1000\ -1000\ -107\ -3]$ Triangular $[-28\ 0\ 22]$ Trapezoidal $[11\ 62\ 1000\ 1000]$
	Angular velocity error e_w	Trapezoidal $[-1000\ -1000\ -73\ -5]$ Triangular $[-30\ 0\ 46]$ Trapezoidal $[1\ 120\ 1000\ 1000]$
	Left torque error T_1	Trapezoidal $[-1000\ -1000\ -110\ -12]$ Triangular $[-39\ 0\ 29]$ Trapezoidal $[11\ 91\ 1000\ 1000]$
	Right torque error T_2	Trapezoidal $[-1000\ -1000\ -117\ -1]$ Triangular $[-48\ 0\ 42]$ Trapezoidal $[7\ 144\ 1000\ 1000]$

Table 6.15 shows the values of the input/output membership functions for the Type-1 FLC found by the chemical reaction algorithm.

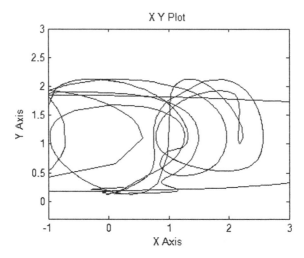

Fig. 6.15 Obtained trajectory of the best result for Case 1, applying the CRA

	Parameters	Case 2
Table 6.12 Simulation parameters and final error of the CRA for Case 2	Individuals	10
	Iterations	15
	Synthesis rate	0.2
	Decomposition rate	0.4
	Substitution rate	0.2
	Double-substitution rate	0.2
	Decomposition factor	2
	k_1, k_2, k_3	36, 328, 88
	Final error	0.077178

Figure 6.21 shows the obtained trajectory of the best result for Case 3. We can observe that even with a smaller error value obtained, the robot was not able to reach the trajectory.

6.2.3 Best Result Applying GAs

In Table 6.16 a comparison of results of the proposed optimization method against other nature inspired techniques that have been applied to solve the same problem is shown [3–5]. The tracking error of the proposed method clearly outperforms the other optimization techniques.

Figure 6.22 shows the best trajectory reached by the mobile when optimizing the input and output membership functions using genetic algorithms.

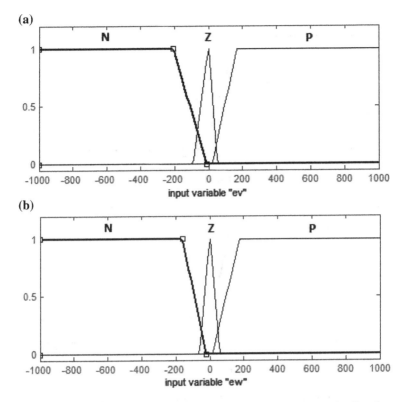

Fig. 6.16 Input membership functions: **a** linear and **b** angular velocity errors for Case 2

Table 6.13 Parameters of the membership functions for Type-1 FLC for Case 2	Variable	Membership function
	Linear velocity error e_v	Trapezoidal $[-1000\ -1000\ 207\ 10]$ Triangular $[-93\ 0\ 51]$ Trapezoidal $[20\ 167\ 1000\ 1000]$
	Angular velocity error e_w	Trapezoidal $[-1000\ -1000\ 160\ 21]$ Triangular $[-62\ 0\ 59]$ Trapezoidal $[14\ 177\ 1000\ 1000]$
	Left torque error T_1	Trapezoidal $[-1000\ -1000\ 213\ 20]$ Triangular $[-50\ 0\ 44]$ Trapezoidal $[20\ 167\ 1000\ 1000]$
	Right torque error T_2	Trapezoidal $[-1000\ -1000\ 157\ 19]$ Triangular $[-47\ 0\ 97]$ Trapezoidal $[16\ 224\ 1000\ 1000]$

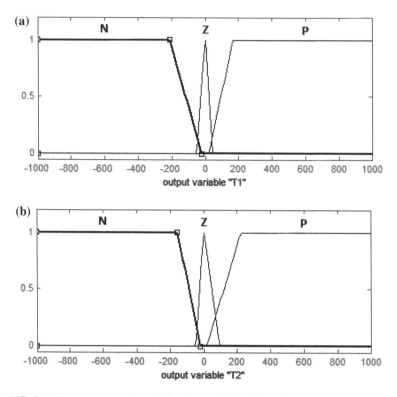

Fig. 6.17 Resulting output membership functions: **a** *left* and **b** *right* torques for Case 2

Table 6.14 Simulation parameters and final error of the CRA for Case 3

Parameters	Case 3
Individuals	10
Iterations	50
Synthesis rate	0.4
Decomposition rate	0.6
Substitution rate	0.3
Double-substitution rate	0.3
Decomposition factor	2
k_1, k_2, k_3	36, 328, 88
Final error	0.061619

6.3 Optimization of a Type-2 Fuzzy Logic Controller

A Type-2 fuzzy logic controller was developed using the parameters of the membership functions found for the FLC of Case 2. The parameters searched with the chemical reaction algorithm were for the footprint of uncertainty (FOU). Table 6.17 shows the parameters used in the simulations.

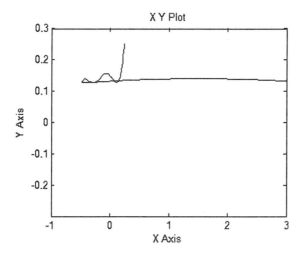

Fig. 6.18 Obtained trajectory of the best result for Case 2, applying the CRA

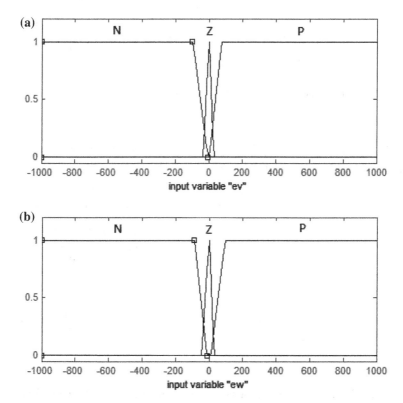

Fig. 6.19 Input membership functions: **a** linear and **b** angular velocity errors for Case 3

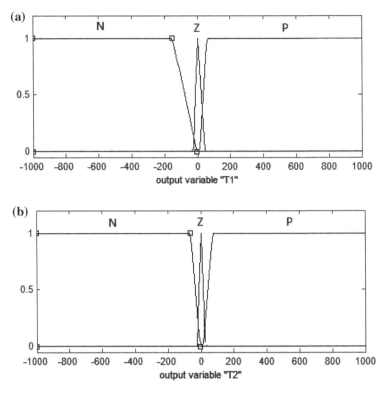

Fig. 6.20 Resulting output membership functions: **a** *left* and **b** *Right* torques for Case 3

	Variable	Membership function
Table 6.15 Parameters of the membership functions for Type-1 FLC for Case 3	Linear velocity error e_v	Trapezoidal $[-1000\ -1000\ -98\ -2]$ Triangular $[-30\ 0\ 27]$ Trapezoidal $[1\ 75\ 1000\ 1000]$
	Angular velocity error e_w	Trapezoidal $[-1000\ -1000\ -89\ -9]$ Triangular $[-42\ 0\ 28]$ Trapezoidal $[9\ 97\ 1000\ 1000]$
	Left torque error T_1	Trapezoidal $[-1000\ -1000\ -152\ -3]$ Triangular $[-22\ 0\ 43]$ Trapezoidal $[9\ 58\ 1000\ 1000]$
	Right torque error T_2	Trapezoidal $[-1000\ -1000\ -61\ -5]$ Triangular $[-21\ 0\ 23]$ Trapezoidal $[9\ 73\ 1000\ 1000]$

Figure 6.23 shows the resulting Type-2 input and output membership functions found by the proposed optimization algorithm.

Fig. 6.21 Obtained trajectory of the best result for Case 3, applying the CRA

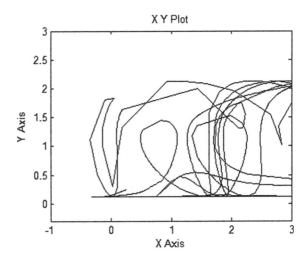

Table 6.16 Comparison of results with different optimization methods for the FLC

Optimization method	Parameters	Average tracking error
Chemical optimization	Number of elements $= 10$	0.0771
	Decomposition rate $= 0.4$	
Genetic algorithm	Crossover $= 0.8$	0.5544
	Mutation $= 0.1$	
	Population $= 100$	
Particle swarm optimization	C1, C2 dynamic	0.1608
Ant colony optimization	Parameters dynamic	0.0866

Figure 6.24 shows the obtained trajectory reached by the mobile robot applying the CRA to the Type-2 FLC.

As observed in Table 6.18, the final error obtained is not smaller that the final error found for the Type-1 FLC. Despite this, the trajectory obtained and showed in Fig. 6.24 is acceptable taking into account that the reference trajectory is a straight line.

In past Figs. 6.15 and 6.21, we observed an "unacceptable" trajectory that was found in the early attempts of optimization for the Type-1 FLC applying this chemical reaction algorithm, in these cases, the parameters found were not the adequate to make the FLC follow the desired trajectory.

In order to test the robustness of the Type-1 and Type-2 FLC, we added an external signal given by Eq. (5.10) showed in Chap. 5.

This represents an external force applied in a period of 10 s to the obtained trajectory that will make the mobile robot to be out of its path.

The idea of adding this disturbance is to measure the errors obtained with the FLC and to test the behavior of the mobile robot under perturbed torques.

Fig. 6.22 Obtained
trajectory using genetic
algorithms

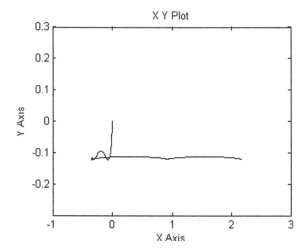

Table 6.17 Simulation
parameters and final error of
the best result applying Gas

Parameters	Value
Elements	10
Trials	10
Selection method	Stochastic universal sampling
k_1, k_2, k_3	36, 328, 88
Final error	2.7736

Table 6.18 shows the parameters for the simulations and the errors obtained
during the run of the simulation.

Figure 6.25 shows the obtained trajectories for the Type-1 FLC optimized with
Genetic Algorithms. From the pictures we can observe that in presence of a disturbance $\varepsilon > 30$, the Type-1 FLC is not able to achieve the tracking control of the
mobile robot.

Figure 6.26 shows the obtained trajectories for the Type-1 FLC optimized with
the chemical reaction algorithm.

Note that the FLC obtained with the chemical reaction algorithm performs better under a disturbance of $\varepsilon \leq 32$.

Figure 6.27 shows the obtained trajectories for the Type-2 FLC optimized with
the chemical reaction algorithm.

In Fig. 6.27 we can observe that with a disturbance of $\varepsilon \leq 34$, the Type-2 FLC
performs better than the Type-1 FLC; the mobile robot is able to return to the
desired trajectory.

With these results, we continued adding a higher level of disturbance (one
unit at the time) to see which level of disturbance is allowed by the Type-2 FLC.
Figure 6.28 shows the obtained trajectories for this experiment.

In Fig. 6.28 we can observe that with a disturbance of $\varepsilon \leq 40$, the Type-2 FLC
is able to keep the tracking, allowing the mobile robot to return to the desired trajectory. For a disturbance of $\varepsilon > 41$, the Type-2 FLC loses control and the mobile
robot is not able to return to the reference trajectory.

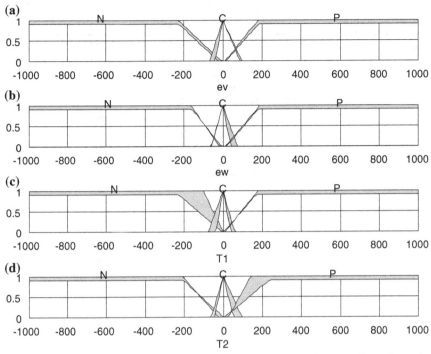

Fig. 6.23 Resulting Type-2 input membership functions, from *top* to *bottom*: **a** linear, **b** angular velocities and output, **c** *right* and **d** *left* torque

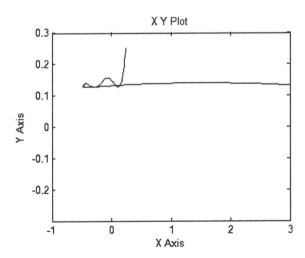

Fig. 6.24 Obtained trajectory for the mobile robot when applying the chemical reaction algorithm to the Type-2 FLC

Table 6.18 Simulation parameters and errors obtained under disturbed torques

ε	Velocity errors	Type-1 (GA)	Type-1 (CRA)	Type-2 (CRA)
0.05	Final error	4.0997	0.9815	29.5115
	Average error	4.1209	1.5823	26.6408
5	Final error	4.1059	0.9729	29.52
	Average error	3.1695	1.8679	26.1646
10	Final error	4.1045	0.9745	29.51
	Average error	3.0985	1.7438	24.9467
30	Final error	4.0912	0.9783	29.51
	Average error	2.2632	1.9481	24.6032
32	Final error	3273	0.9748	29.52
	Average error	3.4667e+003	2.8180	24.6465
34	Final error	1.5705e+004	566.8	29.51
	Average error	1.1180e+004	215.8198	24.9211
40	Final error	2.534e+004	3.5417e+04	29.51
	Average error	186.0611	5.7492e+003	23.8938
41	Final error	8839	3168	685.1
	Average error	2.0268e+004	0.0503e+003	16.5257

Fig. 6.25 Trajectory obtained with the Type-1 FLC optimized with GA's. **a** ε = 30, **b** ε = 32, **c** ε = 34

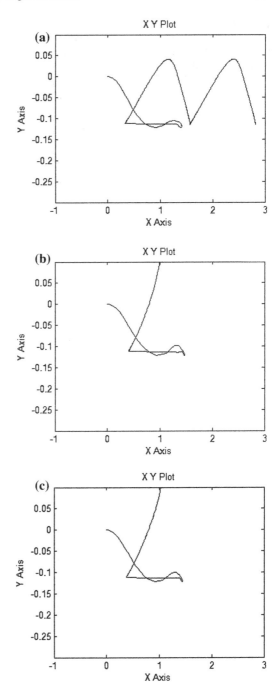

Fig. 6.26 From *left* to *right*,
trajectory obtained with
the Type-1 FLC optimized
with the chemical reaction
algorithm. **a** ε = 30, **b**
ε = 32, **c** ε = 34

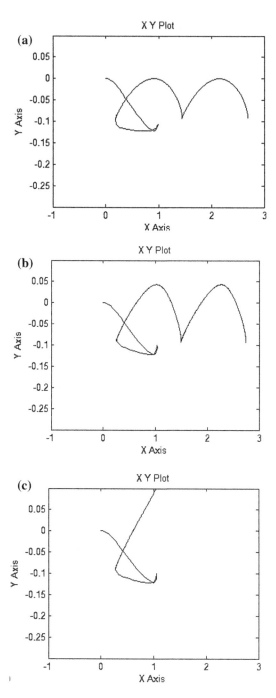

Fig. 6.27 From *left* to *right*, trajectory obtained with the Type-2 FLC optimized with the chemical reaction algorithm. **a** $\varepsilon = 30$, **b** $\varepsilon = 32$, **c** $\varepsilon = 34$

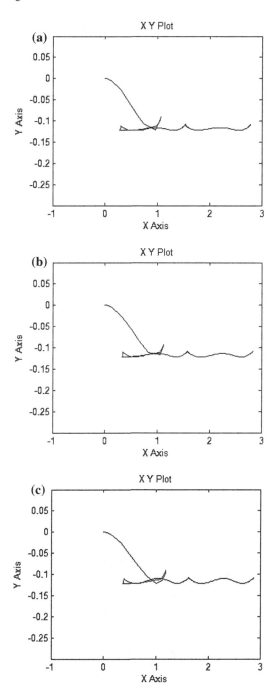

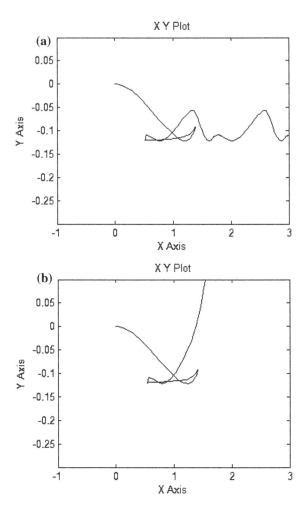

Fig. 6.28 From *left* to *right*, trajectory obtained with the Type-2 FLC optimized with the chemical reaction algorithm. **a** ε = 40, **b** ε = 41

References

1. F. Valdez, P. Melin, Comparative study of PSO and GA for complex mathematical functions. J. Autom. Mob. Robot Intell Syst **2–1**, 43–51 (2008)
2. E. Rashedi, H. Nezamabadi-pour, S. Saryazdi, GSA: a gravitational search algorithm. Inf. Sci. **179**, 2232–2248 (2009)
3. L. Astudillo, O. Castillo, L. Aguilar, Intelligent control for a perturbed autonomous wheeled mobile robot: a type-2 fuzzy logic approach. Nonlinear Stud **14–1**, 37–48 (2007)
4. O. Castillo, R. Martinez-Marroquin, P. Melin, J. Soria, Comparative study of bio-inspired algorithms applied to the optimization of type-1 and type-2 fuzzy controllers for an autonomous mobile robot. Bio-inspired Hybrid Intell. Syst. Image Anal. Pattern Recogn, Stud. Comput. Intell. **256**(2009), 247–262 (2009)
5. R. Martinez, O. Castillo, L. Aguilar, Optimization of type-2 fuzzy logic controllers for a perturbed autonomous wheeled mobile robot using genetic algorithms. Inf. Sci. **179–13**, 2158–2174 (2009)

Chapter 7
Conclusions

Abstract This chapter offers the conclusions about the efficiency and accuracy of the proposed chemical optimization algorithm based on the simulation results for function optimization and fuzzy controller design in robotic applications.

Keywords Chemical optimization algorithm • Type 2 fuzzy logic • Fuzzy control

In this research work, a new optimization method based on mimicking the chemical reactions in nature was introduced. The main characteristics of this algorithm are the exploiting mechanisms combined with the elitist survival strategy, which prevents the algorithm from stagnating in local optima.

The performance of the chemical reaction algorithm (CRA) was evaluated on a set of complex benchmark functions and a type-1 and type-2 fuzzy logic tracking controller. The results were compared with those obtained with another nature inspired paradigms.

For the complex benchmark functions, simulations showed how the algorithm was able to reach smaller values than GAs, PSO and SGA; obtaining good results with a basic set of values; a population of only 10 elements and a maximum of 10 iterations per experiment, except for the Rosenbrock's valley function in which a Genetic Algorithm with a population of 150 individuals and 200 generations obtained better results than the CRA.

The CRA was applied to find the values of the gain constants k_1, k_2, k_3 involved in the equation that achieves the tracking controller. In this stage of the project, the total errors were smaller than the errors reached by means of GAs.

Once the gain constants were obtained, the following stage was to find the membership function parameters used in the fuzzy tracking controller.

When applied the CRA to the fuzzy tracking controller simulations confirmed that the controller found was able to reach the reference trajectory. The algorithm was compared with the results obtained using GAs and again, the result errors were smaller.

L. Astudillo et al., *Chemical Optimization Algorithm for Fuzzy Controller Design,*
SpringerBriefs in Computational Intelligence, DOI: 10.1007/978-3-319-05245-8_7,
© The Author(s) 2014

In the third stage of the fuzzy controller application, the idea was to take the type-1 fuzzy and use it as the primary membership functions. The secondary membership functions were found by means of the CRA and then, a disturbance was applied to the tracking system in order to compare the performance of both controllers (type-1 and type-2).

The error found by the CRA applied to the type-2 fuzzy tracking controller was larger than the errors obtained for the type-1 FLC. Despite this, the trajectory obtained was acceptable.

In order to test the behavior of the type-1 and type-2 FLC's under disturbance, an external force was applied to the tracking system in form of a pulse generator every applied in a period of 10 s to the obtained trajectory; this made the mobile robot to be out of its path.

As resulting of these experiments, the type-2 fuzzy logic controller performed better than the type-1 FLC, for a disturbance of $\varepsilon \leq 40$, the mobile robot was able to return to the reference trajectory. In a previous work [1], these tests were made applying a disturbance of $\varepsilon = 0.03$ maximum.

In this work, the chemical reaction algorithm has proven to be a good optimization paradigm that could be applied to different soft computing problems.

As future work, the parameters used to optimize the Type-2 fuzzy logic controller could vary to obtain a smaller error, this can be achieved by increasing the element pool size and the number of iterations, and a parallel computation can be used, i.e. designing a cluster to divide the task of computing every chemical reaction procedure in separate computers/processors.

Based on the results obtained in these problem applications, we encourage our research community to apply this novel approach to their own optimization problems.

Reference

1. L. Astudillo, O. Castillo, L. Aguilar, Intelligent control for a perturbed autonomous wheeled mobile robot: a type-2 fuzzy logic approach. Nonlinear Stud. **14–1**, 37–48 (2007)

Appendix A

Tables A.1, A.2 and A.3 show the complete results of the chemical reaction algorithm applied to the benchmark functions of Table 5.1 shown in Chap. 5, including the best and worst solutions found by the algorithm, as well as the mean and standard deviation. The simulations were performed 30 times each.

Table A.1 Results of the chemical reaction algorithm (CRA) applied to the benchmark functions f_1 to f_5 of Table 5.1 applying the first set of parameters shown in Table 6.1

Dim	Value	f_1	f_2	f_3	f_4	f_5
2	Best	1.05E-05	2.51E-04	1.69E-04	1.81E-03	1.50E-01
	Worst	1.05E-01	1.18E+01	2.79	5.14E+01	1.67E+01
	Mean	1.07E-02	6.06E-01	2.56E-01	4.69	1.45
	Std	1.76E-02	1.80	4.85E-01	9.82	2.36
4	Best	4.76E-05	1.99E-03	1.11E-05	2.92E-03	2.92
	Worst	2.54	5.04E+01	2.46E+01	1.71E+02	4.00E+01
	Mean	1.94E-01	8.58	4.32	3.19E+01	4.94
	Std	3.97E-01	1.07E+01	5.62	4.00E+01	6.17
30	Best	2.93E-03	7.37E-03	1.42E-02	1.49E-02	2.90E+01
	Worst	6.37E+01	3.70E+03	1.71E+03	1.48E+04	5.25E+02
	Mean	1.75E+01	7.67E+02	3.42E+02	2.64E+03	7.03E+01
	Std	1.82E+01	9.55E+02	4.20E+02	3.47E+03	1.06E+02
32	Best	2.87E-06	1.44E-03	7.28E-04	3.68E-02	3.10E+01
	Worst	9.25E+01	6.64E+03	1.08E+03	1.62E+04	3.33E+02
	Mean	1.67E+01	8.80E+02	2.16E+02	2.55E+03	5.66E+01
	Std	1.96E+01	1.42E+03	2.55E+02	3.36E+03	6.32E+01
64	Best	1.12E-01	6.72E-04	7.00E-02	6.05	6.30E+01
	Worst	2.01E+02	8.66E+03	5.20E+03	4.40E+04	7.02E+02
	Mean	3.19E+01	1.54E+03	6.69E+02	8.66E+03	1.32E+02
	Std	4.71E+01	2.03E+03	1.06E+03	9.70E+03	1.33E+02
128	Best	1.57E-02	7.32E-03	4.96E-02	1.20	1.27E+02
	Worst	2.94E+02	1.98E+04	1.11E+04	1.22E+05	4.95E+02
	Mean	6.77E+01	3.36E+03	1.78E+03	1.91E+04	1.74E+02
	Std	8.29E+01	4.47E+03	2.28E+03	2.61E+04	9.03E+01

(continued)

L. Astudillo et al., *Chemical Optimization Algorithm for Fuzzy Controller Design*, SpringerBriefs in Computational Intelligence, DOI: 10.1007/978-3-319-05245-8, © The Author(s) 2014

Table A.1 (continued)

Dim	Value	f_1	f_2	f_3	f_4	f_5
256	Best	1.09E-04	1.26	2.03E-02	2.00E+01	2.55E+02
	Worst	1.03E+03	4.69E+04	1.46E+04	3.91E+05	4.44E+03
	Mean	2.01E+02	6.55E+03	3.83E+03	4.37E+04	4.91E+02
	Std	2.43E+02	1.01E+04	4.04E+03	7.02E+04	6.41E+02
500	Best	2.88E-01	4.77	5.13E-01	5.65E-01	4.99E+02
	Worst	1.42E+03	1.32E+05	4.48E+04	5.71E+05	2.82E+03
	Mean	3.76E+02	1.87E+04	6.77E+03	7.94E+04	8.00E+02
	Std	4.65E+02	2.74E+04	9.31E+03	1.25E+05	5.65E+02
1000	Best	3.08	1.85	7.53E-03	3.26E+02	9.99E+02
	Worst	3.84E+03	3.09E+05	6.31E+04	9.35E+05	1.99E+04
	Mean	8.07E+02	3.92E+04	1.23E+04	2.05E+05	1.84E+03
	Std	1.08E+03	5.81E+04	1.67E+04	2.59E+05	2.88E+03
1500	Best	1.16E-02	2.45	1.95	6.84E+01	1.50E+03
	Worst	6.69E+03	2.18E+05	1.41E+05	8.52E+05	1.60E+04
	Mean	1.14E+03	5.32E+04	1.68E+04	1.95E+05	2.70E+03
	Std	1.70E+03	5.99E+04	2.47E+04	2.48E+05	2.87E+03

Table A.2 Results of the chemical reaction algorithm (CRA) applied to the benchmark functions f_6 to f_{10} of Table 5.1 applying the first set of parameters shown in Table 6.1

Dim	Value	f_6	f_7	f_8	f_9	f_{10}
2	Best	1.60E-02	−1.56E+03	3.75E-04	1.05E-10	1.47
	Worst	7.70	−2.65E+02	3.84E-01	4.12E-06	2.07
	Mean	1.23	−6.94E+02	5.75E-02	2.56E-07	1.54
	Std	1.52	2.46E+02	7.90E-02	7.66E-07	1.21E-01
4	Best	2.45E-04	−1.94E+03	3.20E-03	1.06E-11	1.16
	Worst	2.24E+01	−5.60E+02	7.36E-01	3.94E-06	2.49
	Mean	7.08	−1.04E+03	2.70E-01	2.33E-07	1.33
	Std	6.19	3.24E+02	1.94E-01	5.75E-07	2.90E-01
30	Best	1.11	−4.61E+03	6.59E-02	2.03E-10	6.03E-03
	Worst	3.01E+02	−1.45E+03	1.92	1.75E-05	1.21E+01
	Mean	1.50E+02	−2.96E+03	9.83E-01	7.66E-07	1.90
	Std	7.92E+01	7.18E+02	3.37E-01	2.56E-06	2.16
32	Best	4.30E-01	−6.96E+03	8.68E-03	2.52E-10	6.16E-02
	Worst	4.82E+02	−1.31E+03	1.45	4.91E-06	7.01
	Mean	1.63E+02	−2.93E+03	1.01	5.82E-07	1.94
	Std	9.46E+01	9.64E+02	3.34E-01	9.72E-07	1.71
64	Best	8.58E-01	−1.05E+04	3.68E-03	2.26E-10	7.52E-02
	Worst	6.65E+02	−2.11E+03	2.70	5.86E-06	1.32E+01
	Mean	3.34E+02	−4.33E+03	1.17	5.51E-07	3.24
	Std	1.73E+02	1.67E+03	5.80E-01	1.13E-06	2.73
128	Best	1.56	−1.37E+04	3.60E-02	8.50E-11	6.84E-02
	Worst	1.44E+03	−3.22E+03	3.82	3.04E-06	1.52E+01
	Mean	6.90E+02	−6.66E+03	1.58	5.01E-07	4.40
	Std	3.61E+02	2.54E+03	8.77E-01	8.63E-07	3.72

(continued)

Table A.2 (continued)

Dim	Value	f_6	f_7	f_8	f_9	f_{10}
256	Best	7.80	−1.57E+04	1.50E-01	8.40E-11	7.20E-02
	Worst	3.45E+03	−4.13E+03	1.00E+01	4.59E-06	1.34E+01
	Mean	1.45E+03	−8.57E+03	2.79	5.14E-07	4.14
	Std	7.61E+02	2.79E+03	2.49	1.05E-06	3.18
500	Best	2.60E-01	−2.56E+04	6.07E-02	2.78E-11	1.38E-02
	Worst	5.10E+03	−5.03E+03	3.59E+01	7.53E-06	7.48
	Mean	2.81E+03	−1.19E+04	4.66	7.49E-07	2.31
	Std	1.36E+03	4.60E+03	5.76	1.67E-06	1.90
1000	Best	6.51	−3.61E+04	7.21E-03	6.05E-12	3.81E-02
	Worst	1.16E+04	−8.61E+03	3.47E+01	8.30E-06	7.47
	Mean	4.89E+03	−1.67E+04	7.48	8.53E-07	1.74
	Std	3.09E+03	5.54E+03	9.03	1.73E-06	1.69
1500	Best	1.87E-01	−5.59E+04	1.81E-02	2.30E-11	9.44E-03
	Worst	1.65E+04	−9.05E+03	6.97E+01	8.68E-06	7.76
	Mean	7.92E+03	−2.04E+04	1.08E+01	7.73E-07	1.51
	Std	4.85E+03	7.94E+03	1.48E+01	1.69E-06	1.62

Table A.3 Results of the chemical reaction algorithm (CRA) applied to the benchmark functions f_1 to f_5 of Table 5.1 applying the second set of parameters shown in Table 6.1

Dim	Value	f_1	f_2	f_3	f_4	f_5
2	Best	4.13E-52	5.17E-51	3.25E-50	4.38E-49	1.33E-02
	Worst	2.83E-40	5.47E-40	1.56E-39	1.82E-39	1.00
	Mean	7.46E-42	3.86E-41	3.27E-41	8.20E-41	7.88E-01
	Std	4.14E-41	1.07E-40	2.20E-40	3.11E-40	3.43E-01
4	Best	1.48E-47	4.15E-45	7.01E-44	2.85E-44	2.66
	Worst	2.79E-36	2.70E-34	1.28E-35	8.46E-34	3.00
	Mean	1.01E-37	8.61E-36	4.38E-37	3.32E-35	2.98
	Std	4.99E-37	4.29E-35	1.90E-36	1.44E-34	4.92E-02
30	Best	8.25E-40	2.43E-38	2.04E-39	2.37E-36	2.88E+01
	Worst	2.36E-31	2.61E-29	8.81E-30	9.47E-28	2.90E+01
	Mean	9.51E-33	6.78E-31	4.98E-31	2.39E-29	2.90E+01
	Std	3.41E-32	3.70E-30	1.48E-30	1.35E-28	2.62E-02
32	Best	3.85E-38	4.26E-37	1.48E-38	1.19E-39	3.10E+01
	Worst	8.68E-30	4.91E-28	6.36E-30	2.24E-28	3.10E+01
	Mean	2.46E-31	1.17E-29	3.85E-31	6.48E-30	3.10E+01
	Std	1.25E-30	6.93E-29	1.18E-30	3.18E-29	7.99E-03
64	Best	8.09E-36	8.98E-36	2.47E-36	3.24E-36	6.30E+01
	Worst	1.43E-29	8.64E-28	8.01E-29	7.63E-27	6.30E+01
	Mean	7.60E-31	4.32E-29	6.60E-30	3.80E-28	6.30E+01
	Std	2.72E-30	1.50E-28	1.60E-29	1.27E-27	1.03E-02
128	Best	4.71E-36	1.14E-35	9.94E-34	3.03E-34	1.27E+02
	Worst	8.29E-29	4.26E-26	2.47E-26	4.40E-26	1.27E+02
	Mean	5.85E-30	1.80E-27	8.22E-28	1.68E-27	1.27E+02
	Std	1.60E-29	7.57E-27	3.60E-27	6.77E-27	1.94E-02

(continued)

Table A.3 (continued)

Dim	Value	f_1	f_2	f_3	f_4	f_5
256	Best	4.75E-35	4.58E-34	1.36E-32	1.66E-31	2.55E+02
	Worst	5.96E-28	2.29E-25	2.80E-25	1.09E-24	2.55E+02
	Mean	2.68E-29	9.17E-27	6.04E-27	2.38E-26	2.55E+02
	Std	9.09E-29	3.59E-26	3.96E-26	1.54E-25	1.50E-02
500	Best	1.01E-32	3.91E-33	2.31E-32	1.43E-31	4.99E+02
	Worst	2.47E-26	8.78E-25	3.70E-25	5.40E-24	4.99E+02
	Mean	9.79E-28	3.41E-26	1.41E-26	2.12E-25	4.99E+02
	Std	3.93E-27	1.52E-25	6.36E-26	8.49E-25	7.33E-03
1000	Best	6.04E-34	2.57E-33	1.54E-32	8.64E-32	9.99E+02
	Worst	5.76E-27	1.10E-24	7.50E-25	3.86E-23	9.99E+02
	Mean	5.95E-28	5.24E-26	3.12E-26	1.09E-24	9.99E+02
	Std	1.30E-27	1.85E-25	1.34E-25	5.47E-24	1.63E-02
1500	Best	4.19E-37	1.68E-31	9.41E-31	7.28E-32	1.50E+03
	Worst	1.53E-25	2.46E-24	4.57E-24	1.37E-22	1.50E+03
	Mean	9.03E-27	1.14E-25	1.31E-25	3.86E-24	1.50E+03
	Std	2.83E-26	3.78E-25	6.52E-25	1.99E-23	1.04E-02

Table A.4 Results of the chemical reaction algorithm (CRA) applied to the benchmark functions f_6 to f_{10} of Table 5.1 applying the second set of parameters shown in Table 6.1

Dim	Value	f_6	f_7	f_8	f_9	f_{10}
2	Best	0	−6.00E+05	0	2.01E-64	1.47
	Worst	0	−4.49E+02	0	7.35E-52	1.47
	Mean	0	−6.80E+04	0	2.30E-53	1.47
	Std	0	1.14E+05	0	1.07E-52	1.35E-15
4	Best	0	−2.07E+05	0	1.07E-65	1.16
	Worst	0	−1.88E+03	0	2.83E-49	1.16
	Mean	0	−3.82E+04	0	5.67E-51	1.16
	Std	0	4.49E+04	0	4.00E-50	8.97E-16
30	Best	0	−1.92E+05	0	7.36E-61	5.01E-07
	Worst	0	−2.50E+03	0	1.06E-50	2.00E-02
	Mean	0	−3.96E+04	0	2.29E-52	1.37E-03
	Std	0	4.19E+04	0	1.49E-51	3.98E-03
32	Best	0	−2.15E+05	0	9.88E-61	1.71E-08
	Worst	0	−3.35E+03	0	2.26E-50	4.08E-02
	Mean	0	−3.52E+04	0	4.76E-52	2.47E-03
	Std	0	4.63E+04	0	3.20E-51	6.97E-03
64	Best	0	−2.54E+05	0	1.68E-61	4.81E-07
	Worst	0	−4.90E+03	0	4.50E-49	1.30E-01
	Mean	0	−6.47E+04	0	9.27E-51	8.22E-03
	Std	0	6.52E+04	0	6.36E-50	2.54E-02
128	Best	0	−1.58E+06	0	8.75E-65	2.96E-07
	Worst	0	−7.76E+03	0	1.40E-51	1.40E-01
	Mean	0	−1.50E+05	0	5.89E-53	1.24E-02
	Std	0	2.75E+05	0	2.54E-52	3.20E-02

(continued)

Table A.4 (continued)

Dim	Value	f_6	f_7	f_8	f_9	f_{10}
256	Best	0	−7.23E+05	0	8.55E-61	1.35E-05
	Worst	0	−9.54E+03	0	1.70E-50	6.48E-01
	Mean	0	−1.19E+05	0	3.50E-52	5.70E-02
	Std	0	1.53E+05	0	2.41E-51	1.31E-01
500	Best	0	−1.05E+06	0	3.81E-63	3.12E-05
	Worst	0	−1.44E+04	0	8.42E-50	1.17
	Mean	0	−1.72E+05	0	1.70E-51	1.56E-01
	Std	0	2.03E+05	0	1.19E-50	2.17E-01
1000	Best	0	−2.23E+06	0	1.54E-60	2.77E-03
	Worst	0	−1.45E+04	0	6.26E-51	9.07E-01
	Mean	0	−2.66E+05	0	1.49E-52	1.60E-01
	Std	0	4.21E+05	0	8.88E-52	1.75E-01
1500	Best	0	−2.97E+06	0	5.76E-60	2.84E-04
	Worst	0	−1.88E+04	0	1.11E-50	8.05E-01
	Mean	0	−4.05E+05	0	2.33E-52	1.53E-01
	Std	0	6.82E+05	0	1.57E-51	1.84E-01

Table A.5 Results of the chemical reaction algorithm (CRA) applied to the benchmark functions f_1 to f_5 of Table 5.1 applying the third set of parameters shown in Table 6.1

Dim	Value	f_1	f_2	f_3	f_4	f_5
2	Best	6.33E-10	1.24E-05	2.82E-03	7.58E-04	8.37E-03
	Worst	3.64E-03	6.67	5.06	1.39E+02	7.40E-01
	Mean	3.16E-04	6.22E-01	6.44E-01	1.08E+01	2.43E-01
	Std	5.99E-04	1.16	1.07	2.10E+01	1.86E-01
4	Best	3.48E-07	3.40E-02	3.08E-02	1.15E-02	2.41
	Worst	3.01E-01	2.25E+02	1.17E+02	1.68E+03	9.46
	Mean	2.90E-02	4.10E+01	2.99E+01	3.10E+02	3.12
	Std	5.63E-02	4.50E+01	3.15E+01	3.33E+02	9.40E-01
30	Best	7.77E-06	2.58E-01	8.34E-02	8.89E-01	2.90E+01
	Worst	1.75E+01	3.27E+04	2.95E+04	4.64E+05	5.22E+01
	Mean	3.13	6.00E+03	9.16E+03	6.18E+04	3.18E+01
	Std	4.67	8.46E+03	9.74E+03	9.42E+04	4.54
32	Best	6.50E-04	1.58E-01	1.03	7.43E-02	3.10E+01
	Worst	1.81E+01	6.67E+04	5.06E+04	5.95E+05	5.41E+01
	Mean	4.00	7.23E+03	1.20E+04	1.22E+05	3.26E+01
	Std	4.53	1.19E+04	1.49E+04	1.41E+05	4.32
64	Best	7.13E-04	2.66E-01	2.40E-01	8.80	6.30E+01
	Worst	1.46E+02	1.92E+05	8.58E+04	2.52E+06	1.02E+02
	Mean	1.41E+01	2.20E+04	2.27E+04	2.58E+05	6.81E+01
	Std	2.49E+01	3.50E+04	2.74E+04	4.69E+05	9.62
128	Best	3.97E-04	1.68E+01	4.76	2.59E+02	1.27E+02
	Worst	2.15E+02	3.41E+05	1.86E+05	3.92E+06	1.89E+02
	Mean	1.75E+01	7.73E+04	3.89E+04	7.96E+05	1.34E+02
	Std	3.57E+01	9.50E+04	4.80E+04	1.05E+06	1.49E+01

(continued)

Table A.5 (continued)

Dim	Value	f_1	f_2	f_3	f_4	f_5
256	Best	1.17E-01	8.69E-01	2.85E+02	2.02E+02	2.55E+02
	Worst	3.83E+02	1.05E+06	7.47E+05	1.13E+07	5.07E+02
	Mean	7.12E+01	1.59E+05	1.27E+05	1.55E+06	2.89E+02
	Std	8.57E+01	2.47E+05	1.74E+05	2.40E+06	5.94E+01
500	Best	2.63E-02	3.65E+02	6.39	1.90E+02	4.99E+02
	Worst	7.10E+02	1.40E+06	1.97E+06	1.65E+07	1.38E+03
	Mean	1.17E+02	2.09E+05	2.38E+05	3.10E+06	5.31E+02
	Std	1.70E+02	3.11E+05	4.03E+05	4.04E+06	1.29E+02
1000	Best	1.13E-01	7.43E+01	3.26	9.72E+02	9.99E+02
	Worst	3.16E+03	2.81E+06	5.07E+06	7.32E+07	1.58E+03
	Mean	3.81E+02	5.51E+05	5.16E+05	7.68E+06	1.05E+03
	Std	5.72E+02	7.41E+05	9.87E+05	1.24E+07	1.12E+02
1500	Best	2.62E-01	4.19E+02	1.10E+02	7.60E+02	1.50E+03
	Worst	3.03E+03	5.40E+06	4.81E+06	8.62E+07	4.34E+03
	Mean	5.11E+02	9.20E+05	7.71E+05	1.07E+07	1.72E+03
	Std	6.81E+02	1.30E+06	1.15E+06	1.75E+07	5.20E+02

Table A.6 Results of the chemical reaction algorithm (CRA) applied to the benchmark functions f_6 to f_{10} of Table 5.1 applying the third set of parameters shown in Table 6.1

Dim	Value	f_6	f_7	f_8	f_9	f_{10}
2	Best	1.37E-04	−1.63E+03	1.21E-06	1.80E-12	1.70
	Worst	4.07E-01	−6.63E+02	2.76E-02	1.33E-08	1.72
	Mean	7.00E-02	−1.04E+03	4.92E-03	1.97E-09	1.70
	Std	9.44E-02	2.09E+02	5.91E-03	2.99E-09	4.66E-03
4	Best	6.99E-04	−2.36E+03	9.49E-05	7.72E-12	1.67
	Worst	5.62	−9.49E+02	2.07E-01	4.55E-08	1.73
	Mean	1.50	−1.53E+03	6.21E-02	5.22E-09	1.68
	Std	1.51	3.45E+02	5.62E-02	8.63E-09	1.25E-02
30	Best	8.84E-01	−6.13E+03	6.42E-04	4.09E-13	1.32
	Worst	1.91E+02	−2.76E+03	1.31	3.91E-08	1.74
	Mean	8.73E+01	−4.31E+03	7.94E-01	5.97E-09	1.37
	Std	5.46E+01	7.82E+02	3.85E-01	8.83E-09	7.36E-02
32	Best	1.24E-02	−7.38E+03	6.46E-06	1.98E-11	1.29
	Worst	2.00E+02	−3.01E+03	1.19	5.91E-08	1.71
	Mean	8.79E+01	−4.24E+03	6.18E-01	9.58E-09	1.34
	Std	5.56E+01	8.67E+02	4.25E-01	1.54E-08	7.90E-02
64	Best	8.57E-03	−1.08E+04	1.22E-04	4.44E-14	6.82E-01
	Worst	4.43E+02	−4.12E+03	1.45	4.38E-08	1.54
	Mean	1.54E+02	−6.10E+03	8.63E-01	7.23E-09	8.14E-01
	Std	1.11E+02	1.42E+03	4.39E-01	9.96E-09	1.97E-01
128	Best	1.34	−1.79E+04	2.55E-03	2.26E-13	1.98E-04
	Worst	9.54E+02	−5.55E+03	2.56	1.74E-07	9.46E-01
	Mean	4.29E+02	−8.76E+03	9.27E-01	1.67E-08	1.03E-01
	Std	2.68E+02	2.18E+03	6.49E-01	3.65E-08	1.87E-01

(continued)

Table A.6 (continued)

Dim	Value	f_6	f_7	f_8	f_9	f_{10}
256	Best	4.23	−1.93E+04	3.80E-06	9.78E-13	1.87E-03
	Worst	2.10E+03	−8.46E+03	5.16	1.82E-07	2.04
	Mean	8.87E+02	−1.23E+04	1.31	1.25E-08	2.66E-01
	Std	6.02E+02	2.04E+03	1.22	2.86E-08	4.23E-01
500	Best	8.22E-01	−3.16E+04	1.06E-04	2.63E-12	7.07E-05
	Worst	4.43E+03	−1.25E+04	1.04E+01	7.18E-08	8.37
	Mean	1.53E+03	−1.73E+04	1.82	9.81E-09	9.99E-01
	Std	1.24E+03	3.60E+03	1.91	1.63E-08	1.59
1000	Best	6.85E-01	−4.05E+04	1.44E-02	2.93E-13	4.87E-03
	Worst	7.75E+03	−1.82E+04	1.97E+01	4.31E-08	1.35E+01
	Mean	4.18E+03	−2.41E+04	4.31	7.65E-09	1.77
	Std	2.06E+03	4.63E+03	5.40	1.06E-08	2.55
1500	Best	5.52	−5.17E+04	7.13E-03	4.36E-13	5.76E-03
	Worst	1.34E+04	−2.24E+04	1.95E+01	9.37E-08	9.44
	Mean	5.58E+03	−3.19E+04	5.12	1.02E-08	1.64
	Std	3.62E+03	6.31E+03	5.79	2.00E-08	2.00

Appendix B

Main Optimization Program for Benchmark Functions

```
RepType='Integer';                              %Representation   Type.
'Binary'/'Decimal','Float','BinFloat'

Nelem = 100;                       %Indicate the number of new elements to
create.

MaxTrials = 10;             %Maximum number of trials to form compounds.

 OutData=outdata;                        %Number of Output Values to evaluate in
Objective Function(Columns OutData x Nelem)

LowerRange=-500;              %Lowest value to be choosen.

UpperRange=500;              %Highest value to be choosen.

Select='ChemSUSOpt';         %Selection Method.

 % BinLenght=4;                      %Lenght of the binary vector (Element)

% SelRate=0.2;                %Number of Elements to be selected to React

RateSynt=0.2;               %Percentage of Elements to be selected to apply
a Synthesis reaction.
RateDeco=0.2;          %Percentage of Elements to be selected to apply a
Decomposition reaction.

RateSubst=0.2;           %Percentage of Elements to be selected to apply a
Substitution reaction.

RateDoubSubst=0.2;      %Percentage of Elements to be selected to apply a
Double-Substitution reaction.

%Step 1: Generate Pool; Create Initial Elements

ElementPool=InitialPoolOpt(LowerRange,UpperRange,Nelem,OutData);


%Initialize Trials

Trial=0;


 %Step 2: Evaluate Initial Elements into a defined problem
```

L. Astudillo et al., *Chemical Optimization Algorithm for Fuzzy Controller Design*,
SpringerBriefs in Computational Intelligence, DOI: 10.1007/978-3-319-05245-8,
© The Author(s) 2014

```
ObjValue = Function7Opt(ElementPool);

BestVal(Trial+1)=min(ObjValue);

minobj=BestVal(Trial+1);

% EvalVal=horzcat(ObjValue,ElementPool);

 %FlagImg=0; %Bandera para salvar la imagen

while Trial < MaxTrials

        if minobj==0, break, end

        %Fitness Ranking Assignment

        RankElem=RankElemOpt(ObjValue);

        %Decomposition Reaction

        Decomposed=DecompositionOpt(ElementPool,RateDeco,RankElem,Select);

        %Synthesis Reaction

        Composed = CompositionOpt(ElementPool,RateSynt,RankElem,Select);

        %Single-Displacement Reaction
        SingleSubst                                                         =
SingleSubstOpt(ElementPool,RateSubst,RankElem,Select);        %Funcion    de
Descomposicion

        %Double-Displacement Reaction
        DoubleSubst                                                         =
DoubleSubstOpt(ElementPool,RateDoubSubst,RankElem,Select);      %Funcion   de
Descomposicion

        OptPool=[Decomposed;Composed;SingleSubst;DoubleSubst];

        ObjValue2       =       Function7Opt(OptPool);        %Evaluar     nuevos
elementos/compuestos

        %Reinsertar nuevos elementos/compuestos manteniendo el tamaño del

        %conjunto inicial

        [SelectNewPool
bestObjVal]=ReinsOpt(ElementPool,OptPool,ObjValue,ObjValue2,Nelem,Select);

        %re-asignacion de variables

        ObjValue=BestObjVal;

        ElementPool=SelectNewPool;

        %Seleccionar mejor elemento/compuesto

        BestVal(Trial+1)=min(ObjValue);


        minobj=BestVal(Trial+1);   %Asignacion   de    variable    solo   para
condicionar el fin del algoritmo.

        %Termina si minobj=0 (el valor cambia dependiendo de la funcion).

        Trial=Trial+1; %Increase Iteration

end
```

Chemical Reaction Sub-Routine Programs

Composition reaction

```
function Composed =
CompositionOpt(ElementPool,RateSynt,RankElem,Select,SyntIndex)

if nargin < 5, SyntIndex = 2; end

    Rate=RateSynt;
    SelectElem=SelectElemOpt(ElementPool,RankElem,Rate,Select);
    [RowWE ColumnWE]=size(SelectElem);

    if RowWE==0 Composed=[];return
    else
        if SyntIndex > RowWE, SyntIndex = RowWE; disp('SyntIndex must not
exceed the number of the selected elements ');end

            if rem(RowWE,2)==0  %Si la cantidad de Elementos en la Matriz
es Par

Composed=SelectElem(1:2:RowWE,1:ColumnWE)+SelectElem(2:2:RowWE,1:ColumnWE);
            else
                Composed1=SelectElem(1:2:RowWE-
1,1:ColumnWE)+SelectElem(2:2:RowWE-1,1:ColumnWE);

Composed2=SelectElem(RowWE,1:ColumnWE)+SelectElem(RowWE,1:ColumnWE);
                Composed=[Composed1;Composed2];
            end
    end
```

Decomposition reaction

```
function
Decomposed=DecompositionOptFIS(ElementPool,RateDeco,RankElem,Select,DecoInd
ex)
Rate=RateDeco;
SelectElem=SelectElemOpt(ElementPool,RankElem,Rate,Select);

    %Decompose Selected Elements
    [A, B]=size(SelectElem);
    if nargin < 5, DecoIndex = 2; end
    NewFiles=A*DecoIndex;
    Decomposed=zeros(NewFiles,B);

    randval=rand;
%    randval=0.5;
    k=0;
    for i=1:A % File Counter
     for j=1:B  % Column Counter

%          Decomposed(i+k,j)=floor(SelectElem(i,j)*randval);
         Decomposed(i+k,j)=(SelectElem(i,j)*randval);
         Decomposed(i+k+1,j)=(SelectElem(i,j)-Decomposed(i+k,j));
     end
     k=k+1;
    end
```

Single Substitution reaction

```
function SingleSubst =
SingleSubstOptFIS(ElementPool,RateSubst,RankElem,Select)
    Rate=RateSubst;
    SelectElem=SelectElemOpt(ElementPool,RankElem,Rate,Select);
    [RowWE ColumnWE]=size(SelectElem);

    randval=rand;
%    randval=0.5;
    if RowWE==0 SingleSubst=[];return
    else
        if rem(RowWE,2)==0  %Si la cantidad de Elementos en la Matriz es
Par
            NewValA=SelectElem(1:2:RowWE,1:ColumnWE)*randval;
            NewValB=SelectElem(1:2:RowWE,1:ColumnWE)-NewValA;
            NewValC=SelectElem(2:2:RowWE,1:ColumnWE)+NewValA;
%            NewValA=floor(SelectElem(1:2:RowWE,1:ColumnWE)*randval);
%            NewValB=floor(SelectElem(1:2:RowWE,1:ColumnWE)-NewValA);
%            NewValC=floor(SelectElem(2:2:RowWE,1:ColumnWE)+NewValA);
            SingleSubst=([NewValC;NewValB]);
        else
%            randval=rand;
            NewValA=SelectElem(1:2:RowWE-1,1:ColumnWE)*randval;
            NewValB=SelectElem(1:2:RowWE-1,1:ColumnWE)-NewValA;
            NewValC=[SelectElem(2:2:RowWE-
1,1:ColumnWE)+NewValA;SelectElem(RowWE,1:ColumnWE)];
%            NewValA=floor(SelectElem(1:2:RowWE-1,1:ColumnWE)*randval);
%            NewValB=floor(SelectElem(1:2:RowWE-1,1:ColumnWE)-NewValA);
%            NewValC=[floor(SelectElem(2:2:RowWE-
1,1:ColumnWE)+NewValA);SelectElem(RowWE,1:ColumnWE)];
%            NewValD0floor(SelectElem(RowWE-1,1:ColumnWE)
            SingleSubst=([NewValC;NewValB]);
        end
    end
```

Double Substitution reaction

```
function DoubleSubst =
DoubleSubstOptFIS(ElementPool,RateDoubSubst,RankElem,Select)
    Rate=RateDoubSubst;
    SelectElem=SelectElemOpt(ElementPool,RankElem,Rate,Select);
    [RowWE ColumnWE]=size(SelectElem);

    randval=rand;
%    randval=0.5;
    if RowWE==0 DoubleSubst=[];return
    else
        if rem(RowWE,2)==0  %Si la cantidad de Elementos en la Matriz es
Par
            NewValA=SelectElem(1:2:RowWE,1:ColumnWE)*randval;
            NewValB=SelectElem(1:2:RowWE,1:ColumnWE)-NewValA;
            NewValC=SelectElem(2:2:RowWE,1:ColumnWE)*randval;
            NewValD=SelectElem(2:2:RowWE,1:ColumnWE)-NewValC;
%            NewValA=floor(SelectElem(1:2:RowWE,1:ColumnWE)*randval);
%            NewValB=floor(SelectElem(1:2:RowWE,1:ColumnWE)-NewValA);
%            NewValC=floor(SelectElem(2:2:RowWE,1:ColumnWE)*randval);
%            NewValD=floor(SelectElem(2:2:RowWE,1:ColumnWE)-NewValC);
            DoubleSubst=([[(NewValB+NewValC);(NewValA+NewValD)]]);
```

```
          else
%                     randval=rand;
                      NewValA=SelectElem(1:2:RowWE-1,1:ColumnWE)*randval;
                      NewValB=SelectElem(1:2:RowWE-1,1:ColumnWE)-NewValA;
                      NewValC=SelectElem(2:2:RowWE-1,1:ColumnWE)*randval;
                      NewValD=SelectElem(2:2:RowWE-1,1:ColumnWE)-NewValC;
%                     NewValA=floor(SelectElem(1:2:RowWE-
1,1:ColumnWE)*randval);
%                     NewValB=floor(SelectElem(1:2:RowWE-1,1:ColumnWE)-
NewValA);
%                     NewValC=floor(SelectElem(2:2:RowWE-
1,1:ColumnWE)*randval);
%                     NewValD=floor(SelectElem(2:2:RowWE-1,1:ColumnWE)-
NewValC);

DoubleSubst=([(NewValB+NewValC);(NewValA+NewValD);SelectElem(RowWE,1:Column
WE)]);
          end
      end
```

Main Optimization Program for k1, k2 and k3 Gain Constants

```
RepType='Integer';              %Representation Type.
Nelem = 10;                     %Indicate the number of new elements to
create.
MaxTrials = 15;                 %Maximum number of trials to form compounds.
OutData=24;                     %Number of Output Values to evaluate in
Select='ChemSUSOpt';            %Selection Method.
RateSynt=0.2;                   %Percentage of Elements to be selected to apply
a Synthesis reaction.
RateDeco=0.4;                   %Percentage of Elements to be selected to
apply a Decomposition reaction.
RateSubst=0.2;                  %Percentage of Elements to be selected to
apply a Substitution reaction.
RateDoubSubst=0.2;              %Percentage of Elements to be selected to
apply a Double-Substitution reaction.


%Step 1: Generate Pool; Create Initial Elements
%Constructor
Constructor = [-300 -30 ...        % 1ra Funcion de Membresia
(trapezoidal), 1ra variable entrada (Negativo)LB
              -100 35 ... % 2da Funcion de Membresia (triangular), 1ra
variable entrada (Cero)LB
               1 101  ... % 3ra Funcion de Membresia (trapezoidal), 1ra
variable entrada (Positivo)LB
              -300 -30 ... % 1ra Funcion de Membresia (trapezoidal), 2da
variable entrada (Negativo)LB
              -100 35  ... % 2da Funcion de Membresia (triangular), 2da


variable entrada (Cero)LB
               1 101 ... % 3ra Funcion de Membresia (trapezoidal), 2da
variable entrada (Positivo)LB
              -300 -30 ... % 1ra Funcion de Membresia (trapezoidal), 1ra
variable Salida (Negativo)LB
              -100 35  ... % 2da Funcion de Membresia (trapezoidal), 1ra
variable Salida (Cero)LB
               1 101  ... % 3ra Funcion de Membresia (trapezoidal), 1ra
variable Salida (Positivo)LB
              -300 -30 ... % 1ra Funcion de Membresia (trapezoidal), 2da
variable Salida (Negativo)LB
              -100 35 ... % 2da Funcion de Membresia (trapezoidal), 2da
variable Salida (Cero)LB
               1 101; ... % 3ra Funcion de Membresia (trapezoidal), 2da
```

```
variable Salida (Positivo)LB

                  -101 -1 ... % 1ra Funcion de Membresia (trapezoidal), 1ra
variable entrada (Negativo) UB
                  -35 100 ... % 2da Funcion de Membresia (triangular), 1ra
variable entrada (Cero)UB
                  30 300  ... % 3ra Funcion de Membresia (trapezoidal), 1ra
variable entrada (Positivo)UB
                  -101 -1 ... % 1ra Funcion de Membresia (trapezoidal), 2da
variable entrada (Negativo)UB
                  -35 100 ... % 2da Funcion de Membresia (triangular), 2da
variable entrada (Cero)UB
                   30 300 ... % 3ra Funcion de Membresia (trapezoidal), 2da
variable entrada (Positivo)UB
                  -101 -1 ... % 1ra Funcion de Membresia (trapezoidal), 1ra
variable Salida (Negativo)UB
                  -35 100 ... % 2da Funcion de Membresia (trapezoidal), 1ra
variable Salida (Cero)UB
                  30 300   ... % 3ra Funcion de Membresia (trapezoidal), 1ra
variable Salida (Positivo)UB
                  -101 -1 ... % 1ra Funcion de Membresia (trapezoidal), 2da
variable Salida (Negativo)UB
                  -35 100 ... % 2da Funcion de Membresia (trapezoidal), 2da
variable Salida (Cero)UB
                  30 300]; ... % 3ra Funcion de Membresia (trapezoidal), 2da
variable Salida (Positivo)UB

ElementPool= InitialPoolFISOpt(Nelem,Constructor);
Trial=0;
ObjValue=10000*ones(Nelem,1);
BestVal(Trial+1)=min(ObjValue);
minobj=BestVal(Trial+1);
while Trial < MaxTrials
%Fitness Ranking Assignment
        RankElem=RankElemOpt(ObjValue);

        %Decomposition Reaction
Decomposed=DecompositionOptFIS(ElementPool,RateDeco,RankElem,Select);

        %Synthesis Reaction
        Composed = CompositionOpt(ElementPool,RateSynt,RankElem,Select);

        %Single-Displacement Reaction
        SingleSubst =
SingleSubstOptFIS(ElementPool,RateSubst,RankElem,Select);   %Funcion de
Descomposicion

        %Double-Displacement Reaction
        DoubleSubst =
DoubleSubstOptFIS(ElementPool,RateDoubSubst,RankElem,Select);   %Funcion de
Descomposicion

        OptPool=[Decomposed;Composed;SingleSubst;DoubleSubst];

        [ROW  COL]= size(ElementPool);
    flag=1;
for i=1:1:ROW

 a=newfis('RobotCHEM');

 a.name= 'RobotCHEM';
 a.type = 'mamdani';
```

```
a.andMethod = 'min';
a.orMethod = 'max';
a.defuzzMethod = 'centroid';
a.impMethod = 'min';
a.aggMethod = 'max';

%Entrada 1, Error de la velocidad lineal
a.input(1).name='ev';
a.input(1).range=[-1000 1000];
a.input(1).mf(1).name='N';
a.input(1).mf(1).type='trapmf';
a.input(1).mf(1).params=[-1000 -1000 ElementPool(i,1) ElementPool(i,2)];

a.input(1).mf(2).name='C';
a.input(1).mf(2).type='trimf';
a.input(1).mf(2).params=[ElementPool(i,3) 0 ElementPool(i,4)];

a.input(1).mf(3).name='P';
a.input(1).mf(3).type='trapmf';
a.input(1).mf(3).params=[ElementPool(i,5) ElementPool(i,6) 1000 1000];

 %Entrada 2, Error de la velocidad angular

a.input(2).name='ew';
a.input(2).range=[-1000 1000];

a.input(2).mf(1).name='N';
a.input(2).mf(1).type='trapmf';
a.input(2).mf(1).params=[-1000 -1000 ElementPool(i,7) ElementPool(i,8)];

a.input(2).mf(2).name='C';
a.input(2).mf(2).type='trimf';
a.input(2).mf(2).params=[ElementPool(i,9) 0 ElementPool(i,10)];

a.input(2).mf(3).name='P';
a.input(2).mf(3).type='trapmf';
a.input(2).mf(3).params=[ElementPool(i,11) ElementPool(i,12) 1000 1000];

 %Salida 1, Torque de Rueda Derecha

a.output(1).name='T1';
a.output(1).range=[-1000 1000];
a.output(1).mf(1).name='N';
a.output(1).mf(1).type='trapmf';
a.output(1).mf(1).params=[-1000 -1000 ElementPool(i,13)
ElementPool(i,14)];

a.output(1).mf(2).name='C';
a.output(1).mf(2).type='trimf';
a.output(1).mf(2).params=[ElementPool(i,15) 0 ElementPool(i,16)];

a.output(1).mf(3).name='P';
a.output(1).mf(3).type='trapmf';
a.output(1).mf(3).params=[ElementPool(i,17) ElementPool(i,18) 1000 1000];

 %Salida 2, Torque de Rueda Izquierda

a.output(2).name='T2';
a.output(2).range=[-1000 1000];
```

```
 a.output(2).mf(1).name='N';
 a.output(2).mf(1).type='trapmf';
 a.output(2).mf(1).params=[-1000 -1000 ElementPool(i,19)
ElementPool(i,20)];

 a.output(2).mf(2).name='C';
 a.output(2).mf(2).type='trimf';
 a.output(2).mf(2).params=[ElementPool(i,21) 0 ElementPool(i,22)];

 a.output(2).mf(3).name='P';
 a.output(2).mf(3).type='trapmf';
 a.output(2).mf(3).params=[ElementPool(i,23) ElementPool(i,24) 1000 1000];


%  Control de activacion de las 9 Reglas del Sistema
  a.rule(1).antecedent=[1 1];
  a.rule(1).consequent=[1 1];
  a.rule(1).weight=1;
  a.rule(1).connection=1;

  a.rule(2).antecedent=[1 2];
  a.rule(2).consequent=[1 2];
  a.rule(2).weight=1;
  a.rule(2).connection=1;

  a.rule(3).antecedent=[1 3];
  a.rule(3).consequent=[1 3];
  a.rule(3).weight=1;
  a.rule(3).connection=1;

  a.rule(4).antecedent=[2 1];
  a.rule(4).consequent=[2 1];
  a.rule(4).weight=1;
  a.rule(4).connection=1;

  a.rule(5).antecedent=[2 2];
  a.rule(5).consequent=[2 2];
  a.rule(5).weight=1;
  a.rule(5).connection=1;

  a.rule(6).antecedent=[2 3];
  a.rule(6).consequent=[2 3];
  a.rule(6).weight=1;

  a.rule(6).connection=1;
  a.rule(7).antecedent=[3 1];
  a.rule(7).consequent=[3 1];
  a.rule(7).weight=1;
  a.rule(7).connection=1;

  a.rule(8).antecedent=[3 2];
  a.rule(8).consequent=[3 2];
  a.rule(8).weight=1;
  a.rule(8).connection=1;

  a.rule(9).antecedent=[3 3];
  a.rule(9).consequent=[3 3];
  a.rule(9).weight=1;
  a.rule(9).connection=1;

      showrule(a);
      individuo=i

     writefis(a,'RobotCHEM13');
     RobotCHEM12=readfis('RobotCHEM13');
```

```
        %Simulacion de la Planta
        sim RobotCHEMTipo1
         %Error en la velocidad lineal
         etotalev1 = length(etotalev);
         etotalev2 = sum(etotalev);  %Suma los valores del error de etotal
         etotalev3 = etotalev2/etotalev1;  %Promedia los valores del error
total

         %Error en la velocidad angular
         etotalew1 = length(etotalew);
         etotalew2 = sum(etotalew);  %Suma los valores del error de etotal
         etotalew3 = etotalew2/etotalew1;  %Promedia los valores del error
total

         eTotal=(etotalev3+etotalew3)/2;
        if eTotal < 0
            eTotal = eTotal * -1;
        end
        eTotal4(i)=eTotal;

        if i~=1
            if eTotal4(i) < eTotal4(i-1)
            writefis(a,'BestRobotCHEM13');
            end
        end
   end
        ObjValue2=eTotal4';
        %Reinsertar nuevos elementos/compuestos manteniendo el tamaño del
        %conjunto inicial
        [SelectNewPool
BestObjVal]=ReinsOpt(ElementPool,OptPool,ObjValue,ObjValue2,Nelem,Select);

        %re-asignacion de variables
        ObjValue=BestObjVal;
%          if flag>2
%              ElementPool= InitialPoolFISOpt(Nelem,Constructor)

%          else
        ElementPool=SelectNewPool;
%          end
        %Seleccionar mejor elemento/compuesto
        BestVal(Trial+1)=min(ObjValue);

        %Estadisticas por Iteracion
        BestPerTrial(Trial+1)=min(ObjValue);
        WorstPerTrial(Trial+1)=max(ObjValue);
        MeanPerTrial(Trial+1)=mean(ObjValue);

        %Estadisticas por Iteracion y Ciclo
        BestsMatrix(Trial+1,Cicle)=BestPerTrial(Trial+1);
        WorstMatrix(Trial+1,Cicle)=WorstPerTrial(Trial+1);
        MeanMatrix(Trial+1,Cicle)=MeanPerTrial(Trial+1);

        minobj=BestVal(Trial+1); %Asignacion de variable solo para
condicionar el fin del algoritmo.

        % Update display and record current best individual
        plot(log10(BestVal),'-k*'); xlabel('Trial'); ylabel('log10(f(x))')
        text(0.5,0.95,['Best = ',
num2str(BestVal(Trial+1))],'Units','normalized');
        text(0.5,0.90,['Worst = ',
num2str(WorstPerTrial(Trial+1))],'Units','normalized');

        drawnow;
        Trial=Trial+1 %Increase Iteration
End
```

Index

L. Astudillo et al., *Chemical Optimization Algorithm for Fuzzy Controller Design,*   77
SpringerBriefs in Computational Intelligence, DOI: 10.1007/978-3-319-05245-8,
© The Author(s) 2014

Printed in the United States
By Bookmasters